PENGUIN BOOKS

CYBERTRENDS

David Brown is a US writer of mixed Native American Indian, Sicilian and Anglo-Saxon descent who now lives in Western Europe. He has travelled widely in Asia, Central and Eastern Europe, the Caribbean and the Americas. A graduate of Columbia University in New York, he was briefly a speech writer and congressional aide on the Washington political scene before turning to full-time reporting in 1980. He was a former correspondent for the *Financial Times* and the *International Herald Tribune*, and European contributing editor of *Business* magazine, and his journalism has been widely syndicated. This is his first book.

CYBERTRENDS

Chaos, Power, and Accountability in the Information Age

David Brown

PENGUIN BOOKS

PENGUIN BOOKS

Published by the Penguin Group
Penguin Books Ltd, 27 Wrights Lane, London W8 5TZ, England
Penguin Putnam Inc., 375 Hudson Street, New York, New York 10014, USA
Penguin Books Australia Ltd, Ringwood, Victoria, Australia
Penguin Books Canada Ltd, 10 Alcorn Avenue, Toronto, Ontario, Canada M4V 3B2
Penguin Books (NZ) Ltd, 182–190 Wairau Road, Auckland 10, New Zealand

Penguin Books Ltd, Registered Offices: Harmondsworth, Middlesex, England

First published by Viking 1997
Published in Penguin Books 1998
10 9 8 7 6 5 4 3 2 1

Printed in England by Clays Ltd, St Ives plc

Contents

1

We Shall Take Us Beyond the Seas

... in that Empire, the Art of Cartography achieved such Perfection that the Map of one single Province occupied the whole of a City, and the Map of the Empire, the whole of a Province. In time, those Disproportionate Maps failed to satisfy and the Schools of Cartography sketched a Map of the Empire which was the size of the Empire and coincided at every point with it. Less Addicted to the Study of Cartography, the Following Generations comprehended that this dilated Map was Useless and, not without impiety, delivered it to the Inclemencies of the Sun and of the Winters. In the Western Deserts there remain piecemeal Ruins of the Map, inhabited by Animals and Beggars. In the entire rest of the Country there is no vestige left of the Geographical Disciplines.

SUÁREZ MIRANDA[1]

In 1629 the Puritan (and self-elected saint) John Winthrop was getting hot under his ruffled white collar. He was mightily annoyed by England's unremitting economic slump as well as by the extortionate taxes that he was expected to pay to his king. Being a man of action, Winthrop resolved to take matters in hand. He found a buyer for his sprawling Suffolk estate, he gathered a band of Puritan adherents around him, and they steered a course over the gray-green waters for the New World. As they neared the promised shore, Winthrop clambered into the forecastle of his ship (the *Arabella*) and addressed a ringing sermon to the would-be colonists gathered on deck. He told them that – as Puritans – they alone had been divinely ordained to establish a model community – a 'city upon a hill' – that would shine for all the world to see. In rising tones, he admonished each to embrace his destiny. They must all keep faith, work hard, and become models of Christian charity. In doing so, they could freely enjoy the ample bounties their God would provide. Winthrop (1588–1649) went on to become the first governor of the Massachusetts Bay Colony and is considered the founder of New England in what eventually became the United States.

A century or so later, when Winthrop's descendants had run through the space they had 'purchased' from the Native Americans of the East, a new rabble of frontiersmen launched yet another westward thrust, which left them sprawled along the California shores. Once again, along the way, they annexed Indian lands, they reaped the windfall riches of a virgin continent, and they further stretched the boundaries of their temporal rule. This latter phase of expansion, which ended only a century ago, was accomplished by these upstart Americans in the implicit belief that they were fulfilling a Manifest Destiny.*

Today, things have changed. America has run out of places to go. The American dream of expansion, however, remains. Indeed, that dream has since been passed on to non-Western cultures as well. Many of those who were once colonized by the West are now beating the colonizers at their own game. But – with all the geographical territory pretty much accounted for – what and where is the next frontier? The leaders of a technically oriented, global can-do culture have been left with no choice but to invent one out of thin air. So, *Welcome to Cyberspace* – the infinite reaches of the Wired West.

*The term Manifest Destiny was coined by John L. O'Sullivan, the editor of the *United States Magazine and Demographic Review* (July–August 1845), when he said that opposition to the US takeover of Texas from Mexico interfered with 'the fulfillment of our manifest destiny to overspread the continent allotted by Providence for the free development of our yearly multiplying millions.' It has since come into general currency, first to describe the view of Americans as a chosen people, divinely ordained to establish a model society on earth, more recently by corporate leaders describing their commercial strategies. For example, Gerald Levin, chairman of the US media conglomerate Time Warner, described dominance in the business of multimedia as 'our Manifest Destiny . . . ' (Mark Landler, 'Time Warner's techie at the top,' *Business Week*, 10 May 1993).

Settle comfortably in your pew and pay heed as today's electronic trailblazers mount the planks and tell of our glorious technical destiny. Watch as the telescreen flickers with seductive visions of the riches and innovative marvels to come. Meet the grizzled outriders who have come from these mythical borderlands to tell of such imminent wonders as *intelligent* buildings; immersive virtual reality; clothing with embedded microelectronic circuitry; a world in which we will each boast a personal camera built into our buttons or hats, all programmed to register our every word and deed automatically. Why, there will even be computers built into the soles of our old blue suede shoes!*

*'MIT's Neil Gerschenfeld uses shoes as computers and the human body as a network. Data could be stored in computer chips in your shoe, and could travel to your fingertips through your body, which can carry a small current. When you shook someone's hand, it would be like making a modem connection: the computers in people's shoes could swap basic information – fax number, e-mail address – with one another.' See James Fallows, 'Navigating the Galaxies,' *Atlantic*, April 1996, p. 106.

These elaborate adult distractions are compelling, to say the least. Anyway, I still get a personal charge whenever I power up my own Apple Mac, and I have been seduced by the potential of the Internet. It's thrilling to introduce unwired friends to the Net – to set them loose with mouse in hand – and then sneak a peek as that light of delight and excitement floods into their eyes. The Net can be a terrific research tool as well. The other day a fellow scribbler rang my doorbell – all bustle and panic – after spending two particularly unproductive hours trying to find a report unavailable at the local library. She was desperate to locate this work as her deadline loomed closer with every tick. We logged onto the Net, revved up a search engine, and snatched the report off the Web in one minute flat.

Benefit all around? Almost – but not quite. There's the flip side to this wizard world as well. Most, by now, have experienced that exquisite water torture known as telephone voice mail – also sometimes called voice *jail* – in which infuriating electronic answering-machines expropriate *our* personal time so that the company we're trying to reach can save the cost of hiring a real telephone operator. It makes me roar with exasperation. It suggests a world of fully interactive (and highly programed) 'choice' where one's options are limited to a combination of touch-tone buttons – and where the possibility of dialog with an actual human being, already a last resort, seems to diminish with each passing day. I see a world in which technical *connectivity* starts to subtly undermine the essentially ethical ties needed to bind and sustain communities, economies, and entire political systems.

Does this help explain why the public fascination with high-tech wizardry is also tinged with undercurrents of ambivalence and intuitive unease? Certainly these undercurrents run much deeper than the fretting surface sea-swells that psychologically accompany an ocean voyage whose outcome can't be foreseen. They are fed by more than blind and unreasoning fear. People intuitively understand that wondrous electronic creations, like all inventions before, also necessarily bring costs as well as benefits. Already it is evident that digital instruments interact with each other in synergistic ways that are unprecedented in the history of machines: since they all inter-operate using the same binary tongue, they can be joined in unpredictable and powerful ways. Indeed, the organic patterns of everyday life are increasingly being 'reconfigured' into a computational mode, as the reader will shortly see. Yet the John Winthrops of our day, those wizened scanners of the digital horizon, behave as if there were few really significant trade-offs or downsides implied. While steering our vessel, they scan the horizon with a disturbingly urgent and obsessive intent. As we embark on a momentous social transition, there is a disturbing absence of frank dialog between the leaders of the technology business, those high priests who enjoy the gift of binary tongues, and the public congregation that absorbs the consequences of their work. This serves to heighten misgivings below decks. Whispers circulate: hints that the digital John Winthrops are confusing their own vested interests with the golden collective destiny that they portray. After all, they are the most immediate and richest beneficiaries of the cybernetic shift.

One point is clear: the digital 'revolution' is being fueled primarily by a drive to create and capture new commercial opportunities. It arrives at a moment when many developed world markets have 'matured' (which is to say that they cannot grow as fast as they previously did), and when their industries confront new challenges from emerging economies abroad. The process of wiring the world is expected to create a fresh set of business relativities. It is thought that the spread of digital systems, combined with the speed at which they operate, will transform entire enterprises at an exponential rate and in the process (hopefully) make them more 'competitive.' It will unleash a process of colossal industrial consolidation and also shift the essential context in which most business proceeds.

The very pace leaves little time to reflect on where this overall procedure might lead. One oracle proposes that where commerce was once predicated on shifting atoms – for example bottles of Evian water or Toyota automobiles (or even computer hardware for that matter) – it is now being entirely reconfigured around the process of transferring digital bits, such as electronic money-market fund transfers, advertising proposals, and coordinated computer-aided design over corporate satellite nets.[2] But the digerati concede that the direction of this shift is unpredictable. They also admit, when pressed, that the transition will be uneasy at first, and that the spread of its benefits will be uneven at best. In fact the process of technological reconfiguration seems to be producing fewer fresh jobs – and creating individual opportunities on a far lower scale – than did the industrial modes that displaced the agriculturally based world.

Most of us who live in developed countries have been directly or indirectly touched. Our economic lives, and those of our friends, are in a state of uncertain flux. For some, adaptations must be made from one month to the next. I know a group of talented photographers who have joined together in a pool – just as the greats who founded the Magnum press agency did in Paris before them. One of these four already enjoys world renown, at least among her peers, for the remarkable strength of the vision she conveys. Rarely attractive to the eye, her images are elusive; their mystery suggests metaphorical rather than literal truths. (As such, they are the perfect antidote to our overliteral times, and thus subtly enhance the vitality of the society in which we live.) Of course, such work is hardly the stuff of commercial success; indeed, this is why she and her associates have pooled together. They fund their artistic work with their earnings as photojournalists. But, whatever their combined creative talent, the economics of their trade seem to be moving against them. The overall cost of kit – the new cameras, the ISDN modems, the processing equipment, and all the rest – is soaring as photography is digitized. The members of the pool take greater personal and financial risks – and their rewards decline. Why? Because the process of making news pictures is now being more seamlessly integrated with the (now) all-digital process of producing magazines and papers. Increasingly, editors can inexpensively download most of the images they need from one of the global pools that enjoy much greater economies of scale and scope. Much cheaper, less waste – and fewer assignments all round for the many smaller players. Only a few – the 'boutiques' – will survive the shakeout to come.

We have all heard the conventional wisdom: that this technically fueled process of 'rationalization' and restructuring will be confined to outdated economic sectors, and to less skilled laborers (who can always be helped to learn a more forward-looking trade). There is little hard evidence to support this hopeful claim. A trained Rover auto employee, once sacked, cannot expect to be re-employed in the high-tech sector by Fujitsu or IBM. In the US, home of the digital elite, the 'industries of the future' employ no more than 2 percent of the working population.

One software strategist notes, 'it may not be too far into the future before law books are computerized and [voice] recognition systems are capable of digesting and assimilating the arguments in a court of law, weighing them against each other, using case history as a point of reference.'[3] But consider: however unpopular lawyers may be in certain quarters, most citizens still embrace the idea that justice should involve a human magistrate and a jury of their peers. Similarly, airline passengers derive a sense of confidence from the real presence of a human pilot at the controls. More and more highly specialized and professional 'functions' – great swaths of the professional class, including lawyering and the piloting of jet aircraft – may nevertheless be next to fall under the digital ax. Their work belongs to the 'costly' working practices, the institutions, and the associated social conventions that evolved to serve the needs of businesses coordinated primarily in physical space. Now, as the operative context of the economy shifts from the physical landscape into a sprawling electronic terrain, and a greater proportion of it is automated by 'smart' machines, the polarities of the labor market are growing more extreme.

The computer-dominated interstices in which markets now have their being operate at such a lightning-fast pace that they powerfully accentuate short-term over long-term priorities. The cycles of battle are so quick that every incremental business advantage can be converted into a profitable market lead. And, if informational tools can deliver this decisive competitive advantage, then the business executive has a fiduciary obligation to his or her shareholders to mobilize these tools, whatever their broader ethical or social costs. This accelerating competitive compulsion explains why so many late-industrial-age traditions of consensus-building are now disintegrating. The Swedish Model, the Japanese *nenko* (or lifetime employment tie), and the German Social Market economy – all reciprocal systems which recognized the interdependence of labor and capital in strong export-oriented economies – are now being added to the list of endangered species.

Interestingly, the computational language in which 'efficiency' is now expressed seems to share a great deal with that of deregulatory, free-market 'liberalism.' Both carefully exclude any serious consideration of the important long-term costs of their agendas, which include the widening inequalities of wealth and opportunity and the resulting social and ecological ferment. Certainly these costs never appear on balance sheets. They remain marginal factors in the digerati's public debate. Only those who are prone by temperament to think beyond the next fiscal quarter seem occasionally to wonder – as did the writer Jerome K. Jerome – how anyone can fully enjoy idling when one hasn't got enough to do. Where will the swelling ranks of the digitally downsized, and those relegated to less lucrative work, find the means actually to buy the wondrous products and services coming down the pike? (Whereas Henry Ford had the common sense to ensure that his employees could buy one of his company's cars, only a small number of wired enterprises – high-tech or otherwise – have adopted a similarly long-term view. For Mexican workers who assemble cars at the Ford plant in Hermosillo for $100 per week, the notion of owning such a car is an impossible dream.) These questions will be thrust upon us with increasing urgency in the years to come, however. The same social and ecological costs that have been provisionally stripped from the profit-and-loss statements of high-tech enterprise will not simply and conveniently go away. As a matter of fact, they are accumulating and multiplying as a stock of unpaid debts. At some point, this will re-emerge to disturb even the wired world's bottom line.

In the interim, global commerce will continue in a boiling and explosive state of flux. How and when it will solidify into new configurations is hard to say. What seems clear, though, is that large portions of the world population will either have to work harder for less or simply find new sources of non-work-related identity. Cyberspace will be an important component in the new layout of life. As a matter of fact, many people may be able to earn decent livings by surfing the Net and taking an active role in cyberspace spectacles. (This idea is elaborated in Chapter 8.) Silicon Valley and Hollywood will soon be rolling out systems that will let us plunge into distractions that are far more engaging than the canned dramas on contemporary TV. For example, consumers will be offered head-mounted, three-dimensional, fully-immersive virtual-reality displays – the personal video Walkman. One is told to expect an explosion of *personal* dream palaces in which, to paraphrase Hunter S. Thompson (the chemically pickled genius of gonzo American journalism), when the going gets weird, the fantasists will *really* get going. Digitized entertainment is quickly emerging as America's fastest-growing and most strategic commercial enterprise. By tapping US software expertise, it is said, people in the near future will be able to interact with each other socially and economically in *virtual* instead of physical space. Communities will cluster at electronic crossroads and saloons. It'll be a new socioeconomic phase – like the Great Depression with a digital twist.

Thus the current, technically induced reconfiguration of the world economy will pose profound concerns not only about the future of most human livelihoods but also about individual and social identity, and about the very future of democracy itself (as the reader will see). So, take heed, brothers and sisters: the future is here. It's high time to give ourselves over to its electronic embrace. Those of us who have been called can only follow with dead certainty and unquestioning *faith*.

Blink – blink.

Is it just me?

Somehow this seems like a tired old drama: this yearning to escape, to reach over the rainbow, to plunge into a glittering realm of golden dreams. Of course the whole extravaganza has been remounted with neon costumes and cyber-trendy props; they suggest that the script, has also been written afresh. In fact it remains unchanged. Indeed, in a hard, contemporary light it comes across as a surreal spectacle, an impoverished morality play that reflects the exhausted ethos of an outdated phase in our hopefully continuing development.

It was in classical Greece that Socrates posed the root question that underlies any ethical discussion, even today: *What constitutes the good life?* This question ought to be paramount at a time when the scenario of continuous expansion and the expectation of ever-increasing material abundance on which it is based, has been exposed as both nostalgic and unworkable: the very antithesis of a truly rational road to any sustainable 'good life' ahead. After centuries of outward campaigns, after rich bodies of inherited wisdom have been lost, after cultural and species diversity have been destroyed, and a profound spiritual emptiness has been settled upon our times, it is clear that the costs of un-limited growth will far outweigh potential future benefits. An evolutionary watershed has arrived, bringing the necessity of uncomfortable choices that many are frightened or unwilling to face. This is a period in which the global village should be exploring alternative visions founded on something nobler and more conservative than a yearning for *more*. Instead, humankind is being asked to harness new technology to the same unimproved goal: to *overthrow* the limits of both time and space. Digerati seek to divert the restive populace with promises of yet another cloud-capped 'city on the hill,' a mythical wonderland where all consumers can pursue a gilded lifestyle based on the insatiable desire to *use* and consume.

My discomfort with the so-called digital 'revolution' stems partly from a sense that it encourages a state of psychological denial: a historical amnesia on the most massive and hazardous scale. Its most vocal advocates routinely ignore or deny the fact that outdated attitudes – like justifying one's expansionary crusades in the name of some religious or technical imperative – have directly contributed to some of the most uncomfortable choices and challenges that we face today. The New Digital World, like many such dream-worlds before, seems to be predicated on avoiding an inescapable confrontation with limits, especially those suggested by the constrained carrying-capacity of a finite earth. I am unsettled, too, by a subtle bias inherent in the language of binary code. Increasingly, public dialog is being purged of the enduring human verities. Equity, love, and compassion; pride in its many shapes; truth, greed, suVering, and sacrifice: the digerati roll their eyes to hear words like these. They irritably explain that – in a future of cyberspace – altogether new 'modalities' will obtain. Let's stop *quibbling*, for heaven's sake – we're floating on an irresistible electronic tide! This is a great technical adventure, an endeavor of evolutionary Darwinism. Only those who sail in the most highly wired arks can hope to prevail.

User! Huddle up to your individual luminescent screen. Witness a preview of the great things to come. Microsoft's founder Bill Gates (author of *The Road Ahead*)[4] will appear in the role of a stern Puritan father promising redemption by technical means. A secret role model of digerati everywhere (if only because he is a self-made plutocrat of staggering personal wealth), he seems exclusively obsessed with 'redemptive' technical dreams. Whenever it appears right and meet for him to do so, he unsheathes his righteous sword, hacks down Luddite and competitor alike, then resolutely cuts and thrusts his way onward, across the illumined landscapes of electronic enterprise, with all the ferocity of an avenging crusader. His achievements speak for themselves. Today most people who use computers gaze at the world through Windows that he has installed on their screens; soon, too, their houses will be wired from the floorboards through the eaves. Meanwhile, Nicholas Negroponte (a Puritan with pomp who authored the book *Being Digital*[5]) goes about his task as if he were sacristan of an Old World church. The director of MIT's renowned Media Lab appears from behind the altar in dramatic vestments and swings a consecrated censer that billows with futuristic digital visions – the entrancing prototypes and 'progressive' concepts that have been so carefully contrived to stir the imagination. By combining a new deregulatory order with their ever-growing cybernetic powers, our church fathers promise to defy the forces of darkness and lead us to the sunny uplands of a fresh prosperity.

And, no mistake, they *have* mounted a breathlessly exciting performance to date. But for those who can filter out the ritual psalms, ignore the sparkling ornaments, and clear the billowing incense from their eyes, a different kind of drama comes to light. It is one based on the timeless tensions between communication and control – a metaphorical struggle between truth and fantasy – an unending contest between the energies of survival and final decline.

Try a simple exercise. Consider the following word:

> **Communicate** [f. Latin *communicat-* ppl. stem of *communica-re* to make common to many, share, impart ...]
>
> *The Oxford English Dictionary*

In this particular definition of the word – one of many – the emphasis is on seeking those elusive commonalities of interest that bind us, one to the next, in a mutually enriching web of diversity.

Now consider this:

> **cy·ber·net·ics** (1948): the science of communication and control theory that is concerned especially with the comparative study of *automatic control systems* (as the nervous system and brain and mechanical–electrical communication systems).
>
> *Webster's Ninth New Collegiate Dictionary*

Note that the emphasis here is not on diversity. It is on *universal* systems, and on the modes and instrumentation of their control.

The word cybernetics (coined by Norbert Wiener, the late MIT professor, in a 1948 book of the same name) is derived from the Greek *kybernetes*, which roughly translated means pilot or governor. This comes in turn from the word *kybernan*, to steer or manage. Although the cassocked digerati who would steer us across the oceans of change promote their cybernetic revolution with a promise that it will *enhance* freedom and communication, they are in fact thoroughly obsessed with control.

Disarmingly, their vision of a global electronic village suggests a soft-edged collection of decentralized networks, all connected by unified protocols (sometimes called 'open systems') that are thoroughly *immune* to authoritarian rule. The resulting totality of the network, the digerati insist, can only take on a spontaneous vitality, generate extraordinary wealth, and spill a cornucopia of beneficent opportunities that will be shared equally by all. To advance this point of view they cite some of the many pleasant serendipities of technological history. They recall, for example, how metalworking emerged. Here was a skill-set that was explicitly harnessed to produce sharper and harder spear points so that warriors could emerge triumphant from battle. But it had unforeseen consequences too: it allowed the fashioning of strong kitchen pots and delicate amulets as well. Similarly pleasant 'second-order benefits' can be expected from digital development. Giant mainframe computers were first built to break secret military codes and to calculate complex missile trajectories. But those great behemoths of sinuous wire and vacuum tubes uncontrollably gave way to powerful and compact personal computers (PCs). Now we are surrounded by a host of miraculous microchip-based tools that make it possible for ordinary people to interact in unexpected and irrepressible ways.

Uncontained, serendipitous progress is the order of the day.

Set against this assessment is an opposing and perhaps somewhat more Machiavellian school of thought. It states that the mere existence of a given technology will not automatically guarantee that its actual or potential benefits will ever be realized. If this were the case, they point out, then universal electronic academies would already have magically appeared to educate all citizens regardless of cost. Clearly they have not. Why? What crucially decides which applications of a given technology eventually *do* become part of everyday life is a special coalition of technological possibility and hard commercial fact. The overall topography of the networked society should therefore follow a businesslike if not wholly predictable pattern – something akin to that which conditions the shape of human settlements, and the relationships of political power, in ordinary geographical terrain. The architecture of our digital cities, the design of our electronic roads and bridges, indeed the priorities and aspirations that reverberate from our tribal drums: all of these will be shaped by an interaction between technologists and moneymen, by the relationship of ordinary consumers to their powerful elites. Just as ports evolved along rivers and within protected bays, and villages around sources of fresh water, the crossroads of electronic commerce will closely coincide with the wellsprings of digital wealth.

With any true revolution, there is an initial period of chaos in which established relativities explode and it seems as if almost any wondrous possibility can take shape. This helps to explain the optimism about (digital) democracy, global enlightenment, and the empowerment of common men and women: similar moods prevailed at the start of the Revolution in France. But canny theatergoers never assume that the pleasure of an enjoyable first act will necessarily be sustained throughout the entire play. In the drama of human change – especially in the worlds of commerce and politics – explosive developments can lead to dark countervailing reactions. Initial chaos gives way to a longer and more involved phase of struggle and consolidation in which ancient relativities are reformulated in fresh forms. New political leaderships replace the old; new economic hierarchies and elites spring to life.

A central focus of the digerati has been the so-called 'information highway' – which is sometimes (mistakenly) equated with the Internet. The Internet is a widely celebrated and physically de-centralized constellation of computers linked to vast stores of

13

information that is accessible to anyone with sufficient resources, computing skills, and tools.[6] Until now, it seems to have exerted a profoundly 'disintermediating' effect on information exchange. Certain kinds of news can now travel faster and bypass the editorial sieve. (For example, a book about a former president of France can circulate the Web despite being banned by authorities in Paris.) After a time, though, many cybernauts start to find themselves drowning in the flood of unsorted data. They demand new editors – new intermediaries – whose essential task (like that of all editors before them) is to filter the (now digital) wheat from the chaff. Software tools – called search engines – arrive on the market to perform this role. How they present data to the user, however, depends almost entirely on the priorities that guided their overall configuration. (This issue will be discussed at greater length in Chapter 4.)

In the revolution now under way, the real question is not, What can the electronics do? but rather, Who will control the keys? Who (or what) will capture the all-important role of trusted intermediary in the digital domains? How transparent will their mediation be? Above all what standards will be used to hold the intermediaries to public account? These questions are hardly limited to the Internet: all of politics, commerce, and society is being networked (and thus reconfigured) along cybernetic lines.

While there are many compelling (and competing) ideas about what the future will bring, there is general consensus on at least one overarching idea: namely, that the networked world will be literally *embedded* with silicon chips. Beneath the blue-green jewel of Earth as seen from space, a sprawling lattice is already taking shape. This invisible substrate of microscopic glass-fiber filaments and pulsing code will shape our daily reality and condition our dreams. As Bill Gates puts it, the economic success of a cybernetically wired society will require 'total participation.' Indeed, its 'pervasiveness is part of the design.'[7]

You can visualize the wired world of the future as a gigantic TV. In your hand, you hold a sophisticated remote-control device that offers you a seemingly endless abundance of choice. Intuitive wisdom suggests that you, as an individual 'user,' will therefore be 'empowered' in the sprawl of networked domains. No longer will you be limited to the fare dished up by a few paternalistic editors of network TV. You can have it *your* way. Yet, of the many options you confront, the most basic, namely the option to tune out, will be denied. What's more, in the world-view of our fully wired fathers, a person's essential individuality can actually be captured in terms of data (and is thus subject to *processing*): it would seem that, for each of us, our very spirit is contained in the DNA that inhabits and 'defines' us. Henceforth, your mind is to be regarded as a complex computational processor – either you will be logged onto the network of life or else you will be dead. The same applies in the economic world. Once we 'users' adopt the informational world-view and successfully boot this new electronic geosphere into life – by purchasing a sufficient number of PCs, mobile phones, electronic belts, and 'smart' blue suede shoes – and when electronic circuitry and code finally and thoroughly saturate the environment in which we exist, then it will not be long before a failure to operate on its terms will be tantamount to economic suicide.

There is still insufficient evidence for the bright hope, so often and casually expressed, that this emerging digital landscape will automatically be organized in democratic ways. Our conceptual apparatus still has to catch up with the fact that the world's ordering principles are migrating away from familiar social codes (like those of constitutional law). Instead, they are being subtly embedded in lines of binary code. Strings of zeros and ones are being used to describe everything from zebras to trees. They will dictate the terms and conditions under which people interact, the way in which business systems inter-operate, the interfaces between people and machines. The values of cybernetics – that towering paradigm of governance by informational means – will determine which segments of society are most empowered in the cybernetic domains. They will define the overall pattern of rights and responsibilities that obtains.

Now, after the long, rich, and frequently bloodstained pageant of human history, a small fraction of our kind live and generally prosper under *elective* governments. Arguably, this stands as one of mankind's most hopeful achievements to date: with democracy in place, people have escaped their fate as playthings of distant forces and unaccountable elites. Of course, a crusty cynic might object that, once the economic shift from an agriculture-based aristocracy to an industrially empowered middle class was under way, the rest was only a matter of time. Without democracy it would have been impossible centrally to coordinate a more mechanical mode of life. But the elevation of democratic principles also demanded shared sacrifice and struggle – a struggle with which millions around the world are still engaged to this day. What's more, the establishment of a more equitable system of political checks and balances has since transcended its economic roots. It has delivered a more nuanced spread of wealth, a higher level of health, generalized education, relative domestic tranquility, and an abundance of civilized arts in the few places where it has firmly taken root. By now, for many of us who enjoy the fruits of this good life, democracy has come to seem like a god-given *right*. How easy it is to forget how exceptional and fragile a privilege it remains, and how vitally it depends on a broad social consensus, an overall climate of mutual trust, and a fundamental ecological equilibrium in which to proceed.

The checks and balances of representative democracy badly need to be upgraded and extended into the emerging cybernetic domains. Today, digital technology is powerfully amplifying the extent to which its most talented and well-funded practitioners can project their distinct agendas and aspirations over society at large. Now, as throughout the great sweep of human history, control of tools will play a decisive role in determining just how far and how effectively individuals, sovereign nations, and powerful commercial enterprises can pursue their own particular dreams. Recall that an Amerindian, armed with arrows, stood precious little chance of repelling the gold-seeking Cortés or the interloping Puritan in his hypocritical white ruff. The invaders imposed their beguiling agenda with fire-spitting cannons, matchlock rifles, and brute blunderbusses.

Today, a similar gap in capabilities is opening out. There will never be an authentic digital revolution so long as a small fraction can access the technology while the overwhelming majority of people remain altogether unaware of its significance. In the spring of 1992, when Los Angeles erupted into fiery riots, many thoughtful observers were dismayed to note that, although one large audio-visual emporium had been utterly scoured of color TVs, the sleek Apple laptops in the adjacent shop had been left untouched. In fact, for all the talk of new frontiers, participation in the goldfields of the Wired West has thus far been limited to a remarkably select few. The vast proportion of the world remains entirely 'unwired' and – judging by current investment trends – will remain so for the foreseeable years to come. (The International Telecommunications Union of Geneva reports that only one in five of the world's population has access to a basic telephone service. There is an even greater disproportion between that *1 percent* of the world who actually enjoy a connection to the Internet, and are happily streaking along the electronic road ahead, and all the rest who lack the means and basic know-how to even locate the access ramps. This should put the information 'revolution' in a different and more sobering light.)

There is another gap opening up as well. When the motive forces of our economic existence are fully shifted into an amorphous electronic matrix, they may prove more impervious to the ethical frameworks, the social obligations, and the traditional forms of regulatory rule built around physical space. Or so, at least, the digerati say; for, while talking about democracy and 'empowerment,' they suggest in the same breath that human networks will increasingly be subordinated to and dependent upon the momentum generated by these machines. The process of technically fueled 'globalization,' however cruel, is supposedly inevitable and cannot be reversed. Indeed, they continue, enlightened futurists must embrace *further* deregulation so as to prosper against competitors in the placeless electronic domains. An attempt to mitigate the extremes of this agenda, for example through higher taxes and public spending, can only result in financial ruin.

The direct result of this agenda is a widening imbalance of wealth and opportunity, a fragmentation of civil society. This is often presented as a split between unwired (and relatively power-less) communities and increasingly powerful globalists (who have mobilized information-based tools). This is too simplistic. The split is not only between wired and unwired tribes, because even those who *do* own modems and PCs lack the blistering mobility and sophisticated connections – coupled with favorable regulatory, economic, and trade concessions – that most empower what some call a new 'overlay culture.' Collectively and figuratively, most of the world's population is being tribalized and its patrimony is being claimed by this colonizing elite; its potential fate begins to look much like that of America's native Indians in John Winthrop's day.

The story of this imbalance is frequently repeated, yet it remains effectively unaddressed. Too little heed has been given to the momentous economic, social, and political consequences of a disequilibrium – fueled by a combination of technology and de-regulation – which may ultimately undermine the digerati's loftiest dreams.

So: is there any reason for the Indians to trust that the Puritans from cyberspace will practice a policy of Christian charity – of live and let live – unless the settlers are inescapably obliged to do just that? None at all. Although it is often claimed that electronic networking has unleashed an 'unstoppable democratizing tide,' its most pro-found effect to date has been to free the newcomers from many tra-ditional forms of democratic accountability. These beneficiaries are reaching for powers that are comparable to those previously enjoyed by the landed aristocrats, the industrial robber barons, and the professional mercenaries who have always been willing to serve both, for a price. Citizens must recapture the political initiative and rebuild the crumbling democratic institutions that were designed to curb the excesses of power in an earlier time and place. The rule of law must be made to apply within the boundaries of silicon and code. Any failure in this respect will rapidly undermine the very foundations on which a sustainable and civilized society rests. The information highway will prove to be a boulevard of broken dreams. Instead of entering a new world, we will find ourselves in a new Dark Age – one that is fractured by rival camps that live in a rumbling and ominous atmosphere of mutual incomprehension, visceral suspicion, and smoldering low-level unrest.

It is tempting but naive to expect a 'revolution' to follow from the mere mobilization of new machines; change requires a parallel and fundamental revamp in the human *context* in which computer systems proceed. Every stakeholder in planetary life must ask the same fundamental (and ethical) question that Socrates posed in his day: namely, What constitutes the good life *in our time*?

Technologists will have to learn to address this issue in tandem with society. Today, when a digital revolutionary mounts the boards, he implies that the real challenges are narrowly technical, and that only experts can participate in the debate. Of course, the greatest single dilemma confronting humankind at this juncture is of a different nature indeed. The developed world wants desperately to maintain its standard of living, the developing world wants nothing so much as to catch up, and the earth simply cannot sustain the combined burdens implied. A global village that is at once more technically interconnected, more economically interdependent, and more ecologically fragile than ever before must above all try to find *a common ethical language* in which to discuss its shared fate; to open its mind to new expectations and dreams; to adopt a more holistic view of progress itself. The current 'revolution,' purportedly built on communication, will be realized only if a crucial distinction is made between qualitative communication and a mere increase in the volume of messages that people exchange. The former is elusive and demands sacrifice – the latter merely requires an extensive installation of costly new machines.

There is a very real danger that the essential search for ethical connectivity will be diverted and drowned in a fractal sea of electronic babble.* In a complex networked world, each individual and every node of the media matrix bears an enormous responsibility to see that the 'revolution' is carried out in its truest sense. The emerging media empires are in a particularly good position to help cultivate

*A note on the use of the word fractal above. Literally, it is defined by the *Shorter Oxford Dictionary* as 'n. A curve having the property that any small part of it, enlarged, has the same statistical character as the whole ... *adj.* Of the nature of a fractal; of ⤳

or relating to fractals.' Fractal geometry, pioneered (and named) by Benoit Mandelbrot in 1975, has become an important mathematical tool in computer modeling of all kinds. Here, however,

I use the word metaphorically to suggest a species of untempered logic that is mobilizing such computing tools and mathematical techniques to better identify structure within evident chaos or complexity – and doing so not merely in the hope of satisfying an innate and irrepressible human curiosity but with the more explicit purpose of subjecting reality to modeling, manipulation, and control. The result of this untempered logic is a fracturing of the whole human and natural landscape.

more informed (and even more constructive) views. Like the financial markets, they have been among the first to deploy advanced information and networking tools to fully globalize their reach. But the jocular moguls who cruise about on oceangoing yachts and soar in private jets from one deal to the next seem to have more quicksilver priorities in mind. Their empires are built on drawing down and diverting the energies on which human survival is based; they befoul the information space with unrealistic images of ever-rising consumption and confuse the tone of public debate. Individuals have passively acquiesced to media(ted) spectacles in which the arts of persuasion, and the need to justify proposals by appeal to the common good, have given way to the dictates of entertainment and to the desire for an instant fix.

Will the Internet change any of this? Certainly it offers cross-roads for lively social contact and debate – and its fragile independence should be celebrated and consolidated at all costs. At the moment, however, for all its ample attractions, the Net remains a hopeful but diverting sideshow. The flood of news coverage lavished upon it, and on the 'information superhighway' with which it is often confused, seems to bear an inverse relationship to its real significance within the wider informational space. Proportionally speaking, the Internet is a minor star in the constellation of networks that surround us and which *together* form the emerging cybernetic domain. It may well grow in size and importance. But the arrival of a wired social space cannot be definitively celebrated until it becomes more obvious how and by whom this space will be managed and mediated. Will adequately informed citizens assert their legitimate claim?

In fact, what many architects of the information age describe as a new world, bristling with 'interactive' (or two-way) networks, is actually evolving in different directions and toward multiple spheres of commercial influence. For instance, there are sophisticated environments of twenty-four-hour 'infotainment,' half fact and half advertising, in which drug makers directly (and of course impartially) 'brief' their sick. Moreover, even as the Internet is being popularized as a vehicle for citizen-to-citizen exchange, technical and financial forces are combining to subtly reconfigure it toward the more familiar and prevalent mass-media mode: namely, one that facilitates heavy 'downstream' traffic (that is, from the media provider to the consumer) but offers only a minimal capacity for 'upstream' traffic (i.e. from consumer to the provider and, more crucially, from one person to the next). This disturbing fact makes it all the more essential that citizens understand the changing nature of networked space. Being like the water we drink, the air that we breathe, and the great natural spaces in which we can roam free, parts of the informational commons can only be conceived in terms of the public good. Thus, if the world is to be redefined as information, and networked across national bounds, then parts of it must never be privatized. It is time to summon the wisdom and foresight to set aside a portion of this territory as a resource to be held and managed as a public trust, just like the great city and national parks of today. (This issue is discussed at greater length in Chapter 7.)

Travelers setting out on great and important enterprises learn that the greatest conquest is that of their own fear. It is important to resist the (easily exploited) instinct to drive impulsively ahead, or to circle the wagons and muffle dissent. By doing so, and by pressing too heedlessly forward, it is all too easy to deplete the very energies that the world will desperately need to weather the transitions to come. The overwhelming challenge posed by the information age is not that of elaborating the means of control but of maintaining creative diversity, and of cultivating the values associated with empathy and consent. The fragmentary language of binary code, built upon a macabre dance of zeros and ones, is simply too narrow to embrace the incomputable human and natural forces that now, as ever before, will decisively drive events. In principle, digital tools can be appropriated to such creative ends. But it is first necessary to acknowledge a profound tension between the configuration of electronically mediated virtual worlds, on the one hand, and, on the other, the eye-to-eye physical and spiritual spaces in which we live. This tension will be resolved only with an understanding that duties and obligations obtain equally in both. The new frontier cannot be entirely free. It must be one in which – to paraphrase the American moral philosopher Reinhold Niebuhr – a life of cyber-rights coexists with cyber-*responsibilities*.*

*His original phrase, coined in 1949 in his book *Faith and History*, was 'Life has no meaning except in terms of responsibility.'

An equally important step is to recognize the extraordinary vulnerability of the electronic vessels to which we have entrusted our future. As we sail across the windy spaces of human transformation, leaving an old world behind and pressing toward the promised shores of a digital dream whose outline cannot quite yet be seen, it is crucial to recall our captivation by a narrow ring of specialists, and our dependence upon powerful but demonstrably fallible computer algorithms – mathematical models of reality. These formulae are expressed in a language that the overwhelming majority of people cannot understand. However, mathematical reality simulations can also embody definite commercial, technical, and political priorities, and these priorities must be far more clearly recognized if the tools themselves are to be managed in publicly beneficial ways.

Cutting-edge scientific studies of chaos and complexity suggest that, in conditions of instability, a small push can redirect the thrust of an enormous enterprise.† This push toward genuine evolutionary progress cannot be delivered by technology alone: it requires an appreciation of the spiritual energies that bind us, and the root values by which we order our lives as men and women. These values take shape in our relationships in natural space, and in the elusive and ineffable realms of heart and soul. So, although the digerati are right to

†'One person or a small group of people acting or applying pressure in the right place at a sensitive moment can have a profound effect on the outcome of a particular situation or, more importantly, on the evolution of wider social relations,' writes Adam Lucas in the journal *Leonardo*. 'In stark contrast to the classical mechanistic [world-view] both chaos and nonequilibrium systems [theorists] maintain that evolution proceeds *because* of the increasing chaos and disorder in the

celebrate the possibility of downloading *War and Peace* from the
Internet (even if it is still uncomfortable to read on the screen), it is
useful to keep a sense of proportion. While leafing through the
pages of an old book in a search
for an altogether different idea, I

universe, not in spite of them ... chaos or
disorder is responsible for spontaneously
generating new order ... this is especially true
with regard to human values ... current social
movements and the values that they embody
will quicken the pace of social change.'
(Adam Lucas, 'Art, science and technology in
an expanded field,' *Leonardo*, vol. 26, no. 4,
pp. 342–4.)

recently stumbled across the following passage, part of a personal
note that Leo Tolstoy penned to his daughters in the early part of
the twentieth century. He wrote:

> ... the views you have acquired about Darwinism, evolution, and
> the struggle for [daily] existence won't explain ... the meanings of
> your life and won't give you guidance in your actions ... [one must
> remember that] a life without an explanation of its meanings ... and
> the unfailing guidance that stems from that explanation ... is a pitiful
> existence indeed.[8]

Tolstoy was a preternaturally inquisitive man. He was deeply in-
trigued by Darwinism as a scientific theory, just as we are fascinated
by the potential of our marvelous new set of machines. However,
curiosity and excitement have never been a sufficient basis for the
conduct of human affairs, and Tolstoy's insight is every bit as apt
now, at the dawn of a new century, as it was in his own troubled
day. Humankind is entering an age of splintering identities, of
radical business restructuring, of bewildering postmodern ethical
relativism. And yet, as contradictory as it may seem, it may also
prove to be an age that wants nothing more than determined constituencies, brave leaders, and a *shared* sense of obligations and
direction as it sails into the tempestuous passages ahead.

2

Chaos Unbound

The earth has been abandoned like a house, it has been decimated ... man
feels a great yearning for space ... to break free from the globe ...

KASIMIR MALEVICH[1]

Switch on the TV news: the chances are good that you will hear a
distant cry of war. You will see chaos, protest, masses swarming this
way and that; a rapid blast of grainy images. The masked profile
of some paramilitary warlord will break into view. Suddenly a
Kalashnikov assault rifle is thrust into the sky. A single round is
squeezed off. Then, in the ominous interval after muzzle blast and
recoil, another terrible rattle bursts forth. It seems almost annuncia-
tory. Through a square, the crowd swarms. A statue topples. The
established government falls.

Amid the violent chaos of confrontations like these, it's no wonder
that video games are such hot property. At least in the digital fantasy
world you – the 'user' – exercise some control. The fighting proceeds according to predictable rules, which can be mastered to
yield a 'win.' In the high-strategy computer game of *Civilization,*
for instance, you are installed in the position of supreme ruler and
invited to 'build an empire that stands the test of time.' And, since
this is one of the more high-brow adult titles, casualties remain dis-
creetly off-screen, appearing as mere statistics on your 'City Display.'
In other entrenched digital fantasies, like Sega's *Mortal Kombat*, kids
are invited to rip off heads and tear out the hearts of their adversaries.

You can lose yourself in a proliferating number of adrenalin-charged alternative worlds that offer 'gameplay so real it's hard to tell where your living room ends and the software begins.'[2] Desert warfare, mystic quests, simulated cities – video games are very big business, eclipsing the total box-office receipts of Hollywood movies.

California artist Matt Mullican has deliberately taken the simulated environment to its most terrifying extreme. One of his works, called *City Project*, is a fifteen-foot computer printout of a 'virtual metropolis' that exists solely in code. It is a vision of sanitized technical perfection, a place from which the more chaotic and incalculable human variables have been scoured. The city's co-ordinates have been stored on a videodisc. You can jack in and let the program take you on a swift race among angular structures and vast deserted boulevards. Spread before you, the simulated electronic buildings are rigidly delineated sections of pure geometry. Neighborhoods are blocked out in flat primary colors – reds, yellows, blues, and greens – perhaps marking inscrutable divisions in society itself. At the center is an oval arena. There are stark rectangular apartment blocks. Monumental avenues lead to sprawling, soundless squares with administrative facilities of suprematist severity. Headlong, you rush past façades embedded with illegible pictograms, shards from a cryptic language of code. There is a massive, bone-white wall, but there are no gates to the outside world (which appears as a mysterious area of blackness at the margins of your screen).*

Mullican's work may lack the virile rush of *Chopper War* or *Cyber-Strike* – games in which you 'move or die, kill or be killed' – but it does capture an essential attribute of all model realities. They are self-contained, predictable: places where the unending confrontation with chaos – human, ecological, political – can be held off for the duration of play.

*Intel chief Andy Grove explains that 'consumers have a *choice*. They can turn on the television or interact with a multimedia PC.' (Louise Kehoe and Paul Taylor, 'Battle for the eyeballs,' *Financial Times*, 23–4 November 1996. (My italics.))

Video games, Matt Mullican's *City Project*: these complex technological simulations powerfully *shape* and enclose our vision of reality.

(*Myst*, an immensely successful interactive game title marketed by Broderbund, advertises itself as a 'surrealistic adventure that will *become* your world.') They are so central to modern life that it seems, from a certain perspective, as if the tools are *creating us* instead of we them: that technology is the prime catalyst behind all evolutionary progress and the answer to every progressive prayer. Remember Arthur C. Clarke's sci-fi classic *2001: A Space Odyssey*? It projects this point of view perfectly. We begin at the dawn of prehistory, when a primitive tribe of half-starved man-apes is introduced by extraterrestrial intelligence to the weapons and technical skills its needs to hunt game, fend off its enemies, and ultimately prevail against the adversities of a harsh, drought-stricken world.

On reflection, though, most of us intuitively realize that our survival is the outcome of a much richer and more complex pageant of human and contextual energies. As *Homo faber*, we have definitely contrived powerful tools. But we are also uniquely social creatures, capable of collective decision and indeed of collective perception. Our *views* shape the course of experience as much as (if not more than) our technical expertise. Moreover, we have the gift of foresight, and can radically reconfigure our cherished assumptions in the light of actual events. We adapt and interact with the earth in a uniquely intelligent and conscious way that implies special responsibilities as well. Any tribe that fails to exploit these advantages, and instead insists on clinging to video-game fantasies when confronting harsh or contradictory facts, will vanish, however excellent its machines.

The relation between survival and outlook is intimate. To illustrate it, I'll conjure an alternative to Clarke's prehistoric scenario. Say you and your hunting tribe have wandered into an unfamiliar stretch of African grassland, wielding the clubs and spears with which you hope to kill game. Naturally, you're aware that many dangers await in an unknown terrain. But your need is acute. Now, experience has conditioned you to look for the greatest threat in the shape of just one creature – call it the wild buffalo. (In your country, lions are virtually unknown.) Yet, as you advance across the savanna, tawny, lion-like outlines seem to take shape in the shadows beneath an island of trees. A few members of the hunt pause to take stock. For the rest, the hunger is too great. Anyway, received wisdom, evolved in a different habitat, maintains that such creatures cannot possibly exist as a threat. Weapons shouldered, the tribe marches straight into the jaws of a crouching pride. Only a bedraggled handful are lucky enough to crawl away with their new (and dearly bought) insight.*

Many generations later, society is similarly stepping forth into a new virtual terrain, armed with a powerful set of tools built around information and its processing techniques. The procession is accompanied by trumpeting sounds in news media that are over-flowing with catchphrases and clichés like 'Digital Revolution,' 'Information Society,' 'Cybernetic Economy,' and all the rest. The idea is that wondrous technologies are creating a new frontier for colonization. Giant markets are being thrown up for grabs. The whole world is being captured in the language of code – and shifted into that alternative (electro)sphere known vaguely as 'the Net'. Suddenly, a voice declaims, there *are* no limits: 'it's finally time to embrace the future with optimism again.'[3]

The time is the late 1950s, and the world is in the grip of the Cold War. In America, a group of technocrats studying the escalating superpower conflict have become convinced that newly elaborated principles of 'game theory' can be combined with emerging techniques of computer simulation and constructively applied to the dangerous nuclear realities of the geopolitical stage. The superpowers might be balanced on a knife-edge of mutually assured destruction (MAD), but (these men argued),

*The tension between entrenched and emerging world-views is continuous in any society or scientific discipline: there is no final equilibrium. In science, MIT's Thomas Kuhn argued persuasively that 'progress,' tradition-ally interpreted as an objective march toward truth, is really an illusion. What actually happens is that scientists adopt a succession of 'paradigms,' intellectual systems that seem, for a time at least, to explain reality best. The earth was observably flat until new intuitions and methods of measurement suggested another perspective. In the interval, the old view was loaded down with so many rationalizations and contradictions that it finally became top-heavy and collapsed. As Kuhn pointed out – much to the irritation of some scientists – the process is essentially nonrational. See his classic *The Structure of Scientific Revolutions* (Chicago: University of Chicago Press, 2nd edn, 1970).

27

given the right information, good algorithms, and sufficiently powerful mainframe computers, the increasingly polarized world could be scientifically managed as a predictable *system*. It could be *modeled*, like Matt Mullican's *City Project*; it could be played, like a giant video game. They said, 'It is information that permits us to take the right action at the right time ... and when information can be quantified and expressed in the symbolism of arithmetic and logic, we have the means to control any real-life situation we can adequately describe.'[4]

The Californian nexus for these Cold War technocrats was the RAND Corporation, a secular monastery where strategy was hatched, elaborately refined, and then presented for debate and eventual adoption in Washington DC. These specialists had the advantage, in uncertain times, of being able to offer absolute certainties. In the cut and thrust of high policymaking, this gave them an edge over the old-fashioned generalists, who were not always so damnably *sure* about things. It is no surprise, then, that the generalists' power and influence waned. It was only a matter of time before the RAND technocrats were running top-dollar Washington departments like Defense. RAND itself became a quasi-official agency. As the prominent US journalist Joseph Kraft wrote at the time, 'the rapidly increasing complexity of war and technology condemns this country to rely on [such] institutions.'[5]

The resulting atmosphere was unreal. Sam Cohen was a nuclear-weapons designer who worked closely with many RAND luminaries: men like John von Neumann, Albert Wohlstetter, Edward Teller, and Hermann Kahn. Cohen recalls von Neumann as a brilliant and charming scientist who was fatally flawed by his addiction to power. When relaxing from his efforts to elaborate game theory, for instance, he studied the feasibility of melting the polar icecaps with nuclear detonations, thereby creating a warmer global climate. It was simply assumed without question that this would be good for all mankind. Meanwhile, in the early 1960s, Hermann Kahn was perfecting the methods of systems analysis. If these methods were applied for just two decades, he predicted, the entire planet would enjoy a standard of living 'equal to or greater than that of the United States.'[6] It was an odd period, in which a heady and assured optimism united with powerful undercurrents of fear. John J. McCloy, the first civilian high commissioner for Germany after the Second World War, was later to say that many at RAND were quite simply drunk on 'the heavy wine of military strategy, methods of destruction, and power politics.'[7] Cohen, also with the benefit of hindsight, describes a group of men who found themselves at the right place at the right time, grasped the unexpected opportunities, and shot straight to the top. The tone of this cabal was set by 'megalomaniacs,' men whose forecasts seemed at times like 'science fiction scenarios ... the product of a lot of imagination ... But they believed, at first honestly and later I think fraudulently, that they had control over this process of recreating the world ... and it all sounded so damn rational and ... reasonable as to be totally unassailable.'[8]

It was the weight of unexpected and incomputable contradictions that brought this elaborate edifice tumbling down. In 1962 the Cuban Missile Crisis illuminated the essential irrationality of decisionmakers acting under conditions of great stress. The Vietnam War punctured more systemic certainties. Vietcong irregulars met the massed technical firepower of the US with low-tech subterranean tunnels, sharpened bamboo bayonets, and a human determination that ultimately (and however improbably) prevailed. In 1965, as the carnage escalated, a RAND theorist inquired of his computer, When will America win the war? The machine instantly replied – *1964*. By its reckoning, the conflict was already won.

When the superpower dichotomy finally passed away with the collapse of the Berlin Wall and the implosion of the Soviet Empire, the now aging Cold Warriors celebrated as if this were a clear triumph of superior Western technology and economic firepower. To an extent it was; yet even the most experienced practitioners of *realpolitik* were caught off guard. More and more, we start to understand, the Communist demise was the result not just of flaws in the Communist system but also of a crisis of legitimacy precipitated by the moral bankruptcy within. Václav Havel speaks of 'the values and principles that communism denied, and in whose name we resisted ... and ultimately brought it down.'[9] He

29

writes that 'communism was not defeated by military force, but by life, by the human spirit, by the resistance of Being and man to manipulation. It was defeated by a revolt of color, authenticity, history in all its variety and individuality against imprisonment within a uniform ideology.'[10]

Many of the cultural and social energies that actually drove these world events were never factored into the policymakers' elaborate equations. And the world of these politicians was more predictable than most, having been divided into satellite states that revolved around one of just two leading ideological suns. The world after the Cold War is more chaotic by far. The TV news is dominated by the angry blaze of fundamentalist religions, reactionary politics, a surge in population growth, and great tides of destabilizing human migration. True globalists, we now pay lip-service to chaos theory and postmodernism. We embrace abstract numerical codes and discount ethical continuities that could bind experience and make our cultural life appear whole. Yet, amid all of this rush of change, surprising continuities remain. As the Cold War technocrats are bowing off the stage, they are being replaced by a new generation of cyberwarriors who sport much hipper clothes and more imaginative coiffures. Their language has also been suitably modified and modernized: instead of talking in terms of clear-cut, linear control, as did their flow-chart-obsessed predecessors at RAND, they now stress *modeling and managing* the behavior of complex systems in flux.[11] Nevertheless, beneath these tokens of apparent change, an underlying outlook, namely the conviction that fundamental human dilemmas can be resolved by inspired technical means, remains largely intact. New tools can help the world's industrial democracies mobilize against a new unitary threat: that of ecological limits. Urgent attempts to understand and *manage* the 'earth system,' so as to extract the last sustainable increments of marginal growth, coincide with inexplicable floods and storms, melting icecaps, and climatic deviations on an unrecorded scale. The new technocrats, who disingenuously portray themselves as the vanguard of a revolution-ary change, respond to these challenges with the mobilization of more elaborately capable tools. In their minds, the world still offers fresh territories for the digital luminary to dominate and *own*.[12]

California: 1995. A rumpled figure bustles distractedly into the room wearing sneakers and a bright floral-print shirt. He is unshaven. His eyes are rimmed from a shortage of sleep and a life spent worshiping before the computer screen. Like so many of his colleagues, Kevin Kelly affects a deliberate informality – a 'lifestyle statement' that thumbs its nose at the blue-suit/spit-polish sartorial style of the old RAND fraternity. Yet, however indifferent in his appearance, this man is part of a remarkably similar clique. An articulate and widely well-read intellectual, Kelly edits *Wired* magazine, a trend-setting monthly that is by turns fascinating and exasperating, and which modestly bills itself as 'the authentic voice of the Digital Generation.' As a founding member of the California-based Global Business Network (GBN), he stands at the inner core of a powerful brains trust that also advises some of the world's largest multinational enterprises, as well as the American presidency, on the influence of information technology and the strategic trendlines shaping the post-Cold War global economy.*

Kelly's magazine paints its vision of the future in a palette of neon-bright, adrenalin-charged tones. *Wired* writes, 'this peaceful, *inevitable* revolution isn't a problem but an opportunity.' It hints at a benign future 'of work outside workplaces, markets without masters, entertainment beyond mass media, civic-mindedness beyond government, community beyond neighborhoods, consciousness that spans the globe.'[13]

*A partial list of GBN's fifty-five major corporate and government clients includes the US president, the Pentagon's joint chiefs of staff, the stock exchanges of London and Mexico, AT&T, most major US media conglomerates, numerous computer groups and advertising agencies, several petrochemical 'majors,' Nissan, Volvo, and the Singapore Ministry of Defense. This information was gleaned from a complimentary article about GBN prepared by one of its own members (Joel Garreau) on commission by *Wired* magazine's editor, himself one of the [new] 'old boys.' See *Wired* (US) 2.11, November 1994, pp. 96–106, 153–8.

Like many others, Kelly is convinced that the world – 'the system as a whole' – is being remade by machines (. . . *and, behold, it is very good*).[14] As Kelly explains, 'what we're doing is making our technology more and more complex, and more biologically lifelike . . .

and as it becomes more useful to us, I believe at some point we're actually going to look at the things we've made and say: *that's beautiful*.'[15] Although Kelly professes to dislike computers, which he calls 'very handy tools,' he *is* attracted by the fact that they are also 'model worlds, small universes. They are ways to recreate civilization.'[16] The cybernetic revolution, he is convinced, is 'absolutely undiluted, pure redemption. It is the most positive force working in the world today.'*

*Mr Kelly does not specify what (or whom) is being redeemed. Presumably the public opinions expressed by GBN members in this chapter are not those of the organization itself. GBN's brief is to offer finely nuanced scenarios that are useful to paying clients seeking to shape realistic corporate investment strategies. (From the *Icon Earth* interview – see note 15.)

Redemption from chaos? Echoes of Genesis! And a flicker of *déjà vu*?

Like the Cold War technocrats from RAND, today's digerati have gained influence in the highest corridors of power. The supposed vanguard of a new global culture, they share three traits, being primarily middle-aged, male, and North American. Well-meaning without doubt, and technical optimists to a man, they also have a strong vested interest in the golden future they portray. They project their vision in uniquely American tones: puritanical perfectionism, a yearning for the open frontier, and an implicit faith that the 'invisible hand of unfettered commerce' will generate the greatest good for the largest number of people. There is also a sense of having to abandon the old world in favor of the new. Here the nationalistic prerogatives of 'Manifest Destiny' come into play.†

†'Cyberspace is the latest *American* frontier,' according to the *Magna Carta for the Knowledge Age* (version 1.2), which speaks of its dedication to renewing 'the spirit of invention and discovery that led ... generations of pioneers to tame the ... continent.'

The *Magna Carta* was prepared under the auspices of the Progress & Freedom Foundation (PFF – 'dedicated to creating a positive vision of the future founded in the historical principles of the American idea'), led by US congressman and House Speaker Newt Gingrich. A cyberspace document posted on

In all of this, the unconverted might scent that recurrent confusion between spiritual redemption and material success; between the goals of a few and those of the many. But America's wagons are driving inexorably across the plains; the mission has been preordained; failure would be tantamount to sin. Their very wagon wheels seem to groan under the weight of this relentless optimism.‡ 'We face the

the Internet in August 1994, it was drafted by Esther Dyson, George Gilder, George Keyworth, and Alvin Toffler, among others. In addition to advocating a strategy of massive government deregulation, the PFF also supports such programs as the colonization of planet Mars.

The PFF is funded by institutions that include AT&T, BellSouth, Cox Cable Communications, Turner Broadcasting Corporation, and *Wired* magazine. The *Magna Carta* is available on the Internet's World Wide Web at http://www.pff.org, or by sending an e-mail request to PFF@aol.com.

‡Of this optimism, this pursuit of an unattainable abstraction called 'happiness,' the French skeptic Jean Baudrillard observes that 'uninterrupted production of positivity has a terrifying consequence. Whereas negativity engenders crisis and critique, hyperbolic positivity for its part engenders catastrophe, for it is incapable of distilling crisis and criticism in homeopathic doses. Any structure that hunts down, expels or exorcises its negative elements risks a catastrophe caused by a thoroughgoing backlash, just as any organism that hunts down and eliminates its germs, bacteria, parasites or other biological antagonists risks metastasis and cancer – in other words, it is threatened by the voracious positivity of its own cells ... ' (Jean Baudrillard, 'The theorem of the accursed share,' in *The Transparency of Evil: Essays on Extreme Phenomena*, trans. James Benedict (New York: Verso, 1993), p. 106.)

21st century [with] an advantage over our foreign competitors. We currently lead the world in computing and communications technologies ... the knowledge revolution ... summons us to renew the dream and enhance the promise [of the] American Idea.'[17] In a settler community that melds the evangelical fervor of a religious revival hall with the headlong frenzy of gold-diggers rushing West, there is little room for dissent. The bottom line? 'Either we're all part of the team or we're not.'[18]

So cyberspace, it seems, is upon us. That this 'city upon the hill' will be largely harmonious and essentially hopeful is self-evident, provided we wholeheartedly embrace the program, promote its further development as a matter of fundamental policy, and unhesitatingly move to 'make change our friend.'[19] For now and the foreseeable time to come, as Voltaire's Dr Pangloss might say, 'everything is for the best in this best of all possible worlds.'*

But it all has a somehow unsettling and familiar ring. The late George W. Ball, a contrarian lawyer, wry statesman, and skilled diplomat known for his unrelenting wit and common sense, once made the telling remark that 'Americans have made a kind of theology out of using science to solve [nontechnical] problems. It's something like a substitute religion; the idea that we can turn to

*Voltaire's Dr Pangloss, in *Candide*, was a fictionalization of Leibniz – that giant of seventeenth-century rationalism who invented binary code. (See Chapter 7.)

these mysterious forces ... that will make us masters of everything ... [and] save us from having to worry about other things.'[20] Norbert Wiener, the father of cybernetics, confined himself to noting how 'power, and the search for power, are unfortunate realities that can assume many garbs.'[21]

It is late autumn in 1990. Night falls over the tarmac at Narita Airport. A flash of helicopter strobe lights punches through smog-choked swirls of rotor wash, and the craft noses toward the pulsing heart of Tokyo. Up ahead, the city is laid out in a vast grid of glowing surface complexity that evokes an overheated microchip, crisscrossed by bright snaking arteries swollen with digital congestion. This is Chiba – a 100 Mile City – an eternity away from Matt Mullican's sterile world and that hopeful future painted by the Cold War technocrats. It is a swarming chaos that has far more in common with Ridley Scott's *Blade Runner* or some dystopian cartoon feature like *Akira* or Mobius's *Long Tomorrow*.* It is staggeringly overpopulated. Traffic is at a virtual halt. Gas masks abound. And the scene here will soon be repeated throughout that uncontained economic explosion that is now Southeast Asia. More people; more affluence; more cars. Which is supposed to be a sign of success, except that there's scarcely room to breathe, much less move.†

Suspended above the fray, the chairman of a leading German business conglomerate, Daimler-Benz, is being swiftly ferried to the first in a series of strategic

*Notes from the cyberculture: Chiba is a down-at-heel industrial quarter in the sprawling Japanese techno-metropolis described by author William Gibson in *Neuromancer*, the novel that gave birth to the term 'cyberspace'. The '100 Mile City' is a phrase coined by *Guardian* columnist Deyan Sudjic in his book of the same name, which explores the millennial force-field of contemporary hyper-urban sprawl. *Blade Runner*, starring Harrison Ford, is director Ridley Scott's fractal vision of persistent humanity in an extrapolated near future of technology run amok. *Akira* is a feature-length animation about a dark secret that lies at the heart of a futuristic Neo-Tokyo, and *The Long Tomorrow* is a dystopian cartoon which portrays the city as a centerless, violent, and inescapable closed loop.

†A recent World Bank study warned that the number of vehicles in Asia is doubling every seven years, that urban air pollution is already at 'critical levels,' and that five out of seven of the most polluted cities in the world are now in that region. See Victor Mallet, 'World Bank presses Asians to protect environment,' *Financial Times*, 6 December 1993.

negotiations with his counterparts at Mitsubishi, a leading Japanese *keiretsu* (or corporate conglomerate). Their intent is jointly to develop and produce automotive guidance systems that will transform the gridlock of city traffic from chaos into a *manageable system*. The basic idea is that each car will be equipped with an electronic 'command center' that strictly governs speed and guides the driver toward the least congested routes. The cars will be centrally controlled via satellite telemetrics. Private vehicles and public transport will be coordinated in a massively integrated traffic-management initiative designed to squeeze a few last increments of capacity out of an overloaded network of roads. In the meantime, captive drivers will be able to choose from a range of interactive car entertainment devices – such as telegambling – to relieve the stress and the boredom of their daily struggle to commute.*

*In mid-1995, the first traffic-management and advanced interactive entertainment systems were being placed on the market in Japan and Western Europe. The Daimler-Benz chairman, Edzard Reuter, was at the same time preparing to retire to his horse-country retreat.

Just a few decades ago, the very notion of in-car navigators had the most glamorous of all possible associations. Moviegoers will recall first seeing this device in an Aston Martin (a very cheetah-like DB5) driven by James Bond in the film *Goldfinger*. Agent 007, played by Sean Connery, uses it to track a nemesis who has devised a mad profitmaking scheme involving the irradiation of a substantial part of the world's gold bullion supply. As Bond's exquisite sports car glides around the serpentine bends of a Swiss alpine road, with the soaring peaks behind, the audience thrills to the erotic charge. The mythmakers have done their work well: Bond's high-tech gadgets (joined of course with the copious lovemaking and jet-setting lifestyle) convey the promise of freedom (or at least mobility) through technical prowess. Instead, as car-navigation technology approaches full realization in the late 1990s, it is switching into a rather different and far less glamorous mode – one that will effectively seal commuting *salarymen* into remotely controlled pods of programmed solitude.

How often the glittering words 'information' and 'revolution' are coupled with a third popular term: 'globalization'. Together, they fuel a belief that 'we are living in a world of one, ever-freer, "modern" and universal ... culture, with an increasingly and beneficially unified political system.'[22] In fact an increasingly wired global economy is also one in which the comfortable lifestyles (and institutions) built around an old order – one which was dominated by the highly developed West – are being undermined by competition from the rising empires of the East. To maintain their rates of growth, the old hegemonic powers are consuming themselves from within: their elites are racing for a stake in the new game while leaving their less agile compatriots to fend for themselves. At the same time, the basic expansionary paradigm is leveling local trades, destroying inherited methods of production, and decimating social relativities and lifestyles associated with both throughout the world. Many digerati look ahead to a global village but ignore the chaotic splintering of identities, the clash of expectations, the deepening resentments and broken dreams that surround them on all sides. The seeds of their global monoculture are being planted in a soil that lacks the essential nutrient of social consensus that they need in order to grow. This consensus will be elusive if not impossible to reach. Why? Because behind the glittering buzzwords lies a more ancient and embarrassing reality. Vermeer, who painted a canvas called *The Astronomer* at the height of the seventeenth-century Age of Exploration, captured it well. He depicts an explorer leaning forward, dreaming of new adventures, measuring the celestial sphere with the palm of his hand. In Raphael's *School of Athens*, the illuminati detachedly toy with multicolored globes: their amusing personal playthings. Even today one is treated to such images as the soft-edged photoportrait of Maurice Saatchi, the wealthy advertising magnate, benignly draping a subtly tailored (and very proprietorial) arm over a sphere marked with the boundaries of his own world empire. There has never been anything accidental about such images. They are deliberately contrived to convey a distinct if subliminal message of patrons whose vision and power encompass the globe: who have measured it, have crossed it, and, most importantly of all, *own* it. This is the way things have always been.

The public is encouraged to revel in the largely insubstantial intoxication surrounding the Net – in which the ability to send e-mail and play video games on a PC is automatically equated with a world cradled in the benign embrace of fiber-optic filaments, digital satellite telephone systems, interactive entertainment, and all the rest. Much less clearly delineated are the decisive commercial motivations that are driving the divisive process of change: these are merely alluded to in articles buried deep inside the business sections of the broadsheet newspapers. Serious analysis is confined to specialized newsletters and the even more esoteric trade press. (How many are even aware of Asia's momentous struggle to develop a satellite production and launch capacity and thus claim a communications niche in the high ground of space? Or of the push to deliver a mobile World Phone Net and 'wire' the markets of 'emerging' economies?) These fragmentary news clips, which better describe the real thrust of 'digital-age' strategy, are like coded shards of an ancient hieroglyph. They have to be painstakingly sought out, reassembled, and finally translated before you can form a coherent picture of the whole. A titanic battle is now being fought for control over the means of information production and distribution in the twenty-first century. It is nothing less than a battle to colonize great swaths of the Wired World: to capture the power that will define a new age.★

To visualize this market more clearly, let's take a quick step back in time: a detour through the rusted remains of smokestack industry. Notice anything peculiar about the old factories? Something in the architectural layout? Executive offices *connect directly* with the old production floors. There was a physical relationship between the means of communication and the layout of production at that time. The only way to make sure that information flowed efficiently between workforce and management was direct: the boss walked down to the factory floor or sent a delegate. So, although those may have been 'bad old days', when bluecollar workers shuffled through a factory gate to punch in at the clock while top brass waltzed through a grand entryway to their waiting cup of coffee, everyone, for good and bad, had to

★For a listing of some of the main players, see *The TeleGeography 100*, ed. Gregory Staple, Jason Kowal, and Zachary Schrag (Washington DC: TeleGeography Inc., 1996). The book divides the industry into 'bit shifters' (like phone network operators), 'bit makers' (like movie studios and software houses), and 'bit processors' (like those who provide network software and hardware infrastructure).

coexist. When, by the turn of the century, telephones made it possible to relay messages over vast distances, an unanticipated consequence was that executives migrated into big-city skyscrapers. What's more, enterprises expanded as the workable scope of their markets spread from local to regional and finally national. A new world was born.

Think of it: *transmission of the human voice*. What once flowed directly between two people over the air that carried their words was now transformed, with the arrival of *a new intermediary*, into an immensely profitable market. By 1912 the telephone industry was the fourth biggest in America.* This was not the outcome of technology per se; it reflected a complex mix of regulatory and commercial relativities (as will be seen). Now, with the digitization of the world, that explosive growth in one industry, telephony, is about to be repeated on a far wider scale. Suddenly it's not just the voice anymore. We are entering a world,

*For a discussion of the phone's trans-formative role in the early-twentieth-century economy, see Ithiel de Sola Pool's fascinating book *Forecasting the Telephone* (Norwood, N.J.: Ablex, 1983), pp. 59–69. As he points out, the evolution of the telephone and its impact on society was crucially shaped by the inextricable combination of technology *and* commercial viability; it was not, in other words, preordained.

it is often said, in which everything is considered to be information: a world in which it is possible to move and manipulate virtually anything that has been expressed in binary bits. Anything that can be captured, converted, and manipulated within the binary embrace potentially becomes the focus of a new market.

Powerful empires will take shape within the virtual communication space that is being interposed between ourselves and nature. Already it powerfully conditions not only our perceptions of the world but also the trajectories of our movement within it. A practical illustration. Say you happen to be the pilot of a modern commercial jetliner. The cockpit in which you fly is a highly sophisticated arcade of distracting lights and multiple screens. The control column has been replaced by an electronic *joystick* functionally identical to those used to manipulate video games. Meanwhile, the network of physical pulleys, cables, and hydraulic actuators that once directly connected your cockpit to the power plant and flight surfaces has also been scrapped. As a matter of fact, so has *any* direct interaction with the engine and air currents. So you've been edged altogether out of 'the feedback loop.' You're interacting with numbers in cyberspace. Your perception of the flight depends on what the intermediary (in this case a digital telemetric system) flashes onto your screen. These systems *measure* coordinates, *simulate* natural feedback, and *convey* your instructions to the plane. The craft's ultimate safety depends on the extent to which the software-generated virtual world corresponds and responds to physical reality.

This trend can also be seen on the fractal landscape of contemporary warfare – military *and* commercial. (After all, what is contemporary business life if not a form of sublimated battle?) Defense of the realm is now carried out by electronic proxy.★ Victory crucially depends on dominating the flow of information.†
This explains why the Allies' first objective in the 1991 Gulf War against Saddam Hussein was to cripple Iraqi communications: without electronic eyes and ears, the renegade was tactically and strategically blind. Meanwhile, both the Allied protagonists and the general public back home experienced the reality of war at an unusual remove: it was seen from the vantage of video cameras mounted in the nose cones of guided missiles. The actual mechanics of death-dealing were both distanced and sanitized through the use of bizarre euphemisms, stripped of moral content, whereby 'bombing a place flat was "servicing a target," or "visiting a site"'[23] and where the overall spectacle – billed as 'the

★Of course, the history of warfare has always been one of increasing distance between the protagonists. In the days of Greece and Rome, battle was a bloody, hand-to-hand affair which caught up everyone from the lowliest regular to the highest general officer. Arrows, catapults, rifles and missiles steadily increased the distance between enemies to the point where, today, they most often fight without ever coming face to face.

†In July 1996 Deputy US Attorney General Jamie Gorelick told a Senate subcommittee that the possibility of 'an electronic Pearl Harbor' is a very real danger for the US. She noted in her testimony that the US information infrastructure is a hybrid public/private network, and warned that electronic attacks 'can disable or disrupt the provision of services just as readily as – if not more than – a well-placed bomb.' At the same hearings, subcommittee members learned that about 250,000 intrusions into Defense Department computer systems are attempted each year, with a 65 percent success rate. The Clinton administration official was quoted in the *BNA Daily Report for Executives*, 17 July 1996, cited by Edupage editors on the Internet, 21 July 1996. See Edupage archives at *http://www.educom.edu/*.

showdown in the Gulf' – had as much in common with entertainment and television ratings as with the defense of a new world order.★

★The nether-world of skullduggery is being reconfigured cybernetically as well. The apparatus that was once mobilized for military espionage has been redirected in the post-Cold War world toward maintaining and enhancing national corporate competitiveness. Intelligence agencies like the National Security Agency, the CIA, and MI5 (understandably anxious to reinvent a relevant role for themselves and justify their swollen budgets) treat journalists and political insiders to seductive hints about the secretive battles now being fought daily by 'console jockeys' in the fluctuating electronic byways of cyberspace. For further information and references, see 'Codes, Keys, and Conflicts: Issues in US Crypto Policy.' ACM Report (*Communications of the ACM*), June 1994.

As with phones and jets and cybernetic battle management, so with the economy as a whole. Cybernetic webs of production are being established as the central characteristic of the world economy.[24] In principle, they yield greater flexibility and efficiency, and lower cost. The practical result is that the workforce must be redefined as another of many 'inputs'

✦ Of the estimated $28 billion US intelligence budget, roughly $25 billion is used to fund satellite and electronic espionage. Only rarely do such activities see the light of day – as in early 1995, when France expelled a team of five CIA agents who are alleged to have been conducting a program of espionage into the French telecommunications and electronic industries. The US government has countered that French agencies had targeted over seventy major US corporations and finance houses, including Boeing, IBM, Texas Instruments, and Corning Glass. The CIA also estimates that 80 percent of Japan's espionage effort is economic, and that it is mainly directed against the US. It would seem the Cold War has been replaced by a technological arms race, and that the cloak and dagger have been mothballed in favor of the cyberdeck. Moreover, the spread of powerful surveillance equipment is not limited to government and big business: such equipment is increasingly a consumer commodity as well. Within the catalogs of Spycatcher and Lorraine Electronics one finds a breathtaking assortment of electronic exotica, freely available on the open market: everything from a manifestly absurd 'bug' concealed in an electronic cocktail olive to a more sinister unmarked van delivered complete with the most advanced surveillance equipment money can buy. See Michael Cassell, 'Bugs infest the boardroom,' *Financial Times*, 16 January 1993.

✦ On the corporate side, retinal eyescans, voiceprint identifiers, mass genetic testing, and other intrusive technologies signal the increasingly chaotic and fast-changing context of corporate security. In the 1970s this was primarily still built around protecting bricks and mortar – the physical plant and equipment of the industrial enterprise. Now a company's most valuable resources are in human or digital form. They reside in computerized product prototypes, proprietary manufacturing processes, the accumulated know-how of its employees. These offer ready targets to rivals. For instance, British Airways was recently obliged to pay a settlement to its competitor Virgin after admitting that 'rogue' employees had hacked into Virgin's database for information about customers, load factors, and pricing schemes.

In *Count Zero*, one of the famous 'cyberspace trilogy' written in 1986, author William Gibson portrays a highly paid corporate technocrat forced to hire a mercenary force to engineer his escape to another firm. That vision is not as surreal as it seems. Arthur D. Little, the management and high-tech consultancy, went to a US court in 1994 to get an injunction to prohibit its General Motors-owned competitor EDS from soliciting Little employees *anywhere* in the world. This came after EDS had lured away virtually the entire staff of Little's aerospace consultancy, resulting in a significant loss of business. Companies are increasingly putting contractual restrictions on

that feed a fragmented series of electronically dispersed (and often independent) subsidiaries, frequently referred to as 'nodes'.★ Workers, unskilled and skilled, are confronted with two formidable competitors: people elsewhere and systems deliberately designed to supplant their tasks.[25]† Indeed, the technology is destroying more jobs than it generates, and nowhere is this as pronounced as in that 'engine of future economic growth': sunrise high-tech enterprise.‡ Result? The income gap between the rich and poor widens, job tenures shorten or disappear, and insecurity mounts. Simply put, it now takes fewer people to develop, build, and sell a product than ever before. The very notion of full-time work is becoming 'a social artifact that has outlived its usefulness';[26] one-third of the world's working population, numbering some 820 million, is already either

employees, aimed at preventing them from taking their 'trade secrets' and putting them to work in the service of a competing firm. Rival carmakers Volkswagen and GM have been involved in a protracted corporate-espionage dispute that has drawn in the governments in Washington and Bonn at the highest levels. It was triggered by the defection to VW of a senior GM executive, the Spaniard José Ignacio López de Arriortúa, together with a good part of his expert staff. They are alleged to have brought with them secret computer files detailing GM's strategic plans.

★The word 'node' literally means a point around which subsidiary parts are organized. Here, used in its cybernetic sense, it implies an operating entity that revolves around one computer in a sphere of networked machines.

†' ... the automatic machine ... is the precise economic equivalent of slave labor,' Norbert Wiener wrote. 'Any labor which competes with slave labor must accept the economic conditions of slave labor.' (*The Human Use of Human Beings* (Garden City, NY: Doubleday, 1954), p. 162.)

‡Between 1987 and 1991, employment in the IT sector in the US, which accounts for the largest proportion of such jobs within the OECD, fell by 18.9 percent. During the same period in Germany it fell by 5.8 percent, and in Italy by 16.5 percent. For the whole of the OECD it fell by 8.3 percent between 1985 and 1987, and was stagnant for the period between 1987 and 1991. (Vivian Bayar and Pierre Montagnier, 'The Information Technology Industry,' *The OECD Observer*, no. 198, February/March 1996, p. 39.)

jobless or working longer for a wage that is insufficient to cover costs.*

*This is the highest reported level since the Great Depression of the 1930s. See the 1994 annual report of the International Labor Organization, Geneva. It remains to be seen whether the information economy will deliver new kinds of jobs as it takes root, much as the Industrial Revolution ultimately created new forms of livelihood after a period of uncomfortable transition. To date, it seems to be absorbing people at a much lower rate. Given the scale and scope of the current upheaval, moreover, a more immediately pressing question is whether societies can weather the transition with their democratic institutions intact. This is discussed later.

(Uncharitable cynics might conclude that if there *has* been a cybernetic Renaissance, it is best sought in the sinuous world of managerial euphemism, where employees of a company undergoing a 'downsizing' might find themselves 'surplus to requirements' and thus politely offered 'the opportunity to seek a function elsewhere.' In other words, their 'careers have been superseded'; which is to say, they have been 'deselected' in the process of 'human re-engineering.' The result, of course, is that they face 'involuntary termination.' In the lexicon of enterprises that are programed to 'rationalize,' there is no place for terms that might admit to actual responsibility.)

Instead of replacing large-scale industrial production, the cybernetic economy is revolutionizing its methods, globally redistributing the allocation of work, and, above all, shifting its motive forces into a new informational space. In the process, it is also introducing a nervous gap between winners and losers, and, according to Ethan Kapstein, director of studies at the Council on Foreign Relations in New York, 'leaving millions of disaffected workers in its train ... inequality, unemployment and endemic poverty have become its handmaidens.'[27]

Such a split has never before occurred on such an all-encompassing scale. However, it is not completely without precedent. When telephone technology first appeared on the scene, it accelerated a shift away from old locally rooted businesses and family farms toward an economy of sprawling national business conglomerates. It took time for people's sensibilities and social institutions to catch up. Workforces were left fragmented and unable to immediately organize and mount an effective response. As a matter of fact, it was not until well into the 1930s, at the height of the Great Depression, that labor associations in the US won the legal right to collectively bargain with their employers on equitable terms. (In the interval, union organizers were routinely roughed up and murdered by hired thugs. Miners shouldered weapons to fight lawmen. Some strikers actually seized industrial facilities, while many others sought refuge in the great dream palaces of Hollywood entertainment.) This violent and unsettled period, with high unemployment and mounting social insecurity, seems to have been largely forgotten by many in the developed world today. It was nevertheless a time that posed the greatest challenge that democratic society has hitherto faced. Policymakers in Franklin D. Roosevelt's New Deal administration of 1932–6 recall their sense of facing a momentous choice of either adapting to the new relativities or presiding over a collapse; either organizing an orderly revolution or witnessing a downfall of the system itself. Europe meanwhile smoldered with the flames of fascism and militarism that eventually ignited the Second World War.

The irony is that all of this turbulence was fed by people's un-fulfilled wish not to defeat the system so much as more fully to *engage* in its revised economic and social narratives. This experience remains relevant in light of the trajectory upon which we've been launched today. The pillars of stability and social consensus that eased the initial strife of industrialization – universal suffrage, trade unions, welfare states and all the rest – are crumbling away. As the Net opens the way for an electronically wired global market economy, and as new empires built on exploiting a radically expanded definition of 'information' emerge, citizens will be called upon to make quick conceptual leaps and devise new social and political means of managing the unsustainable excesses that are already taking shape. They will have to redefine the increasingly meaningless concepts of political 'left' and 'right'; they will have to make difficult choices between 'openness' and self-protection; they will have to build constituencies for a more balanced sharing of costs and benefits within their own borders. In many cases, too, they will have to bear the costs of higher social spending and accept the need for more flexible and effective regulatory regimes. This is admittedly a tall order. But lacking such initiatives, and considering the unprecedented pace at which the current transition is proceeding, the danger is that the world will once again hear the angry clash of history repeating itself in an even more violent and destabilizing shape.

But enough of all this exasperating history. Let's jack back into the bright 'lite' world of video games and computing entertainment. The realm of nifty software packages like DivorceX – a perfect tool if you happen to have favorite holiday snapshots that include former husbands or wives whom you'd rather forget. Hey – no problem! Fire up the old deck, snip out the offending spouse, and paste in your current paramour instead. The result? Digitally composited reality. How delightfully convenient to confound the real with the virtual!

Of course, Hollywood pioneered the 'morphing' technique for moving pictures. In *In the Line of Fire*, Clint Eastwood is a Secret Service agent who has been given a chance to redeem an early career snafu. In a horrifying flashback sequence, we look on helplessly as he rides shotgun on the fender of JFK's limousine. The snafu was the terrible assassination in Dallas, and it unfolds in a seamless and utterly plausible composite of temporally disconnected film 'footage.' Director Steven Spielberg's *Jurassic Park* dinosaurs were modeled, animated, rendered, and composited by digital means: the film was a hybrid mixture of real life and complete simulation. Finally, in the 1992 film *Lawnmower Man*, directed by Brett Leonard, we attain the cloud-capped heights of morphing nirvana. Whole stretches of the film sprawl through an utterly virtual space, an all-enveloping sensory experience with no material existence whatever (apart from the electrons streaking through the labyrinths of microcircuitry and code).

The cybernetic future being promoted by many contemporary policymakers and pundits is a similarly misleading collage. It contains bits of truth which have been subtly decontextualized to yield a plausible yet insubstantial dream. Throughout the brief and remarkable history of the cybernetic 'paradigm,' its downsides have been consistently unanticipated, misunderstood, or simply ignored. So, wise hunters of the postmodern tribes, take care. Even the light-filled savannas hide treacherous patches of shade. Fire up your modem (if you have one), log onto the Internet (if you can afford it), and download the *Magna Carta for the Knowledge Age* (assuming you belong to that upper one percentile of the world's population that actually knows how). Although the main objective of this *Magna Carta* is to advocate an agenda of thoroughgoing deregulation of the communications industry (which is incidentally described as its 'Manifest Destiny') and to scold as patriarchal and 'socially elitist' those who advocate a more measured approach, you will find that this portentous document also ever-so-modestly proposes that 'the central event of the 20th century is the overthrow of matter.' In fact 'wealth in the form of physical resources is losing its significance ... the central resource is ... knowledge.'[28] Where humankind once had its historical eras dominated by bronze, iron, and oil, we are now streaking at warp speed into a future in which the fusion of information and high technology will be the 'primary resource for generating economic prosperity.'[29] Even the Europeans in Brussels have managed to cobble together a consensus on this: 'the move toward an information society ... will in the long run be as important as the first industrial revolution.'[30]

Those who absorb the costs of this transition — not only individuals and enterprises in the West but also that unwired majority of the world (40 per cent of whose population still till the soil) — have yet to take the full implications on board. They have been assured that a pervasively networked world will automatically be more benign and preferable to the one in which they already happen to live. It is as if the mere existence of a global Net will confer widespread enlightenment, a fair distribution of resources, and general 'empowerment' to the many and various societies of the world. Using new tools — satellites, sensors, massively parallel computers — mankind will miraculously comprehend and manage the world 'system' in ever more refined and intelligent ways. A modest dose of 'appropriate technology' in the right space in good time will dependably yield a luxuriant flowering of individual growth, intellectual development, and the progress of more humanitarian ideals.

However, when you look closely at almost any single component of this so-called digital economy, the picture becomes more confused.* Apart from the high costs associated with owning and managing computer networks, there is also profound disagreement about what actually constitutes an informational property right – an issue of fundamental importance to any 'information-based' society. Do you 'own' your telephone number? How about electronic records of your buying patterns? Your medical history? And who should be responsible for protecting your rights and enforcing new laws in this field? '*Information*,' as the author Stewart Brand famously pronounced, '*wants to be free*.'[31] Using digital computers, it is possible to reproduce things more easily and distribute them faster than ever before. Anne Branscom, a Harvard-based communications lawyer, suggests that 'the ease with which

*The basic outline of an information economy was laid out by Daniel Bell in *The Coming of Post-Industrial Society* (New York: Basic Books, 1973). Evidence of its existence is both mixed and particularly complicated by the lack of any reliable method of measurement. A few random facts:

✦ Over a century ago, in 1870, two-thirds of US gross domestic product was accounted for by manufacturing of material commodities, with services comprising the rest. In 1990 the proportions were reversed. Information technology was one of many factors in this structural shift; but of course the rise of automotive mobility and associated businesses was probably more crucial.

✦ In 1994 world trade in manufactured *goods* powered ahead by 9 percent – the fastest rate of growth in over two decades. Much of this growth was generated by non-fuel commodities and by communication hardware, according to the World Trade Organization in Geneva. The value of cross-border services meanwhile slowed in both real and relative terms. It is interesting to note that, of the world's six largest economies, the two worst trade performers overall during the 1980s (the US and the UK) were the most dependent upon high tech. The strongest players (Japan, Germany, and Italy) did best with a mixture of high, medium, and low tech.

✦ In 1994 a study by McKinsey, the international management consultants, failed to find any company whose success could be attributed primarily to the superior use of high tech. It concluded that information technology has mainly speeded the 'delayering' (i.e. elimination) of middle management, and an increase in the volume of information flows (particularly in finance).

✦ The notion of communications as a commercial 'service' barely existed before the telephone. With the advent of newer and related systems (including television, telefax, satellites, modems and so forth), coupled with progressive miniaturization and increasing cheapness and power of microcircuitry, the cost of processing and distributing information of all kinds has plunged. Hence, the related growth in the supply and velocity of information being moved. If this is the information revolution, then its outcome will still depend on controlling supply and trade routes. It is precisely this battle that is now under way.

electronic impulses can be manipulated, modified and erased is hostile to the deliberate legal system that arose in the era of tangible things.' Data that were once protected, precisely because they had a physical shape, or because they were difficult to reproduce and/or move – or indeed because, like the voice, they were not perceived as 'data' at all – can now 'be compromised in ways invisible to the human eye and at speeds almost unintelligible to the human mind.'[32]

The emerging system is clearly more complex. But this stems primarily from the fact that high-speed data communication tools have created a potentially boundary-less global space without the basic rules needed to ensure its stability. Countervailing regulatory powers are still fragmented along national lines. For example, valuable databases on the buying habits, medical records, incomes, and personal histories of private individuals are surging into radically underregulated offshore data havens, just as tax-allergic fortunes have streamed for years into cashbox companies 'based' in Switzerland, Liechtenstein, and the palm-fringed Antilles. But, while technology has made it easier for this to happen, it is the lagging *legal* framework that has created incentives for this chaos-inducing behaviour to take place. This does not mean regulation is impossible. In fact, once information is shifted over into a network, for example, it is trivially *easy* for companies with adequate resources to manage the movement and protect the value of their own intellectual property in cyberspace. For example, they will be able electronically to identify their customers and 'tag' data to ensure that they are properly paid. Will ordinary citizens be in the position to exert similar control? In whose interests will the informational world be managed?

This question requires value judgments which are obscured by the convenient myth that 'the machine' is hopelessly spinning 'out of control.' To explain the course of current events, the emerging elite borrow liberally from chaos theory, which looks at how systems, such as fluids in a turbulent state, evolve from flux. They suggest that an increasingly complex world economy is now most usefully perceived as an ant colony or a beehive: that is to say, as a 'self-regulating system.' If problems exist – and they do – these should not be seen as the outgrowth of inequitable policies, or of

47

imbalanced technical and/or economic trajectories, but rather of misguided efforts to correct them with an excess of 'unwieldy rules.' Confronted by 'unstoppable technical tides,' the only rational stance for a public policy regulator is one of laissez-faire.* Indeed, some digerati even assert that representative government has become 'the last great redoubt of bureaucratic power on the face of the planet.'[33]† The family of man is far more usefully understood as 'a collection of unthinking automata in a vast hive mind.'[34] One guru of the digital age – MIT's Media Lab chief Nicholas Negroponte – admires the lyrical V-shape formations made by migratory ducks. Their supposed *absence of leadership* strikes him as a most agreeable feature. Peering at the birdlife in nature's blue sky he sees 'a highly responsive collection of *processors* behaving individually and following simple harmonious rules without a conductor.'[35]

However, as world business takes on more of the attributes of the machines that run it, there seems to be a widening divide between the agendas that are computationally possible and more measured approaches that are humanly sustainable. It is unlikely that the transition to a virtual economy will be as smooth – for most of the world's population – as its privileged technical protagonists suggest. Even Adam Smith, who vividly (if inaccurately) likened the industrial world to a machine, accepted the need for valves to vent steam and *bound* chaos. But the mind-set of the cybernetic elite, being computational, refuses to make even this basic provision: the need to balance potential cyber-dominant late-twentieth-century economic orthodoxy – that of anarchic individualism battling to dismantle the civic infrastructure and escape its regulatory 'heavy hand.' Of course, human societies have tended to bound, or regulate, chaos in different ways. For example, there is broad popular support for centralized regulatory control over petrochemical plants, nuclear reactors, and ballistic missiles, to name just a few rather complex systems. Moreover, systemic theories tend to ignore that irrational and transcendent gift that is unique to man: the occasional capacity to unite and share sacrifice in pursuit of a higher ideal.

*Ironically, this very *expectation* powerfully conditions the flow of actual events.

†In his classic *Inquiry into the Nature and Causes of The Wealth of Nations* (1776), Adam Smith proposed that markets are best regulated by 'the invisible hand' of market forces. In this century his position was elaborated in the economic theories of Friedrich von Hayek, who proposed that the more complex a society becomes, the less subject it is to central control. This is being combined with many of the insights put forward by chaos theory to create the

netic efficiencies with human restraint. Like the video game, their minds work in a closed loop. Their value system refers only to self-serving and quantitative priorities, and it deals with the more incalculable human energies by trying, quite simply, to keep them at bay. The fantastic 'networked growth' scenarios take little account of long-term social, ecological, or political viability: that is to say, the degree to which the high costs of competitive 'retooling' and continuous economic integration can be sustained in the everyday world. The truth is that humankind faces a crisis of transition of monumental and possibly unprecedented proportions. Change – the surest sign of life – is now taking on a radically *discontinuous* quality. For most of history it evolved within a framework of more settled relativities. For instance, if a storm swept through a coastal fishery, then a grandfather, father, and son could take stock, rebuild the boats, and mend their nets. The fishing life would go on. Now the fishery has been exhausted by sonar-equipped factory ships, the family has been dispersed, and the community and habits built around the local economy have faded away. Prevailing relativities change with the blink of an eye. Commerce is driven by ever more capable and faster technology. Data fly about and change course at the speed of light. A company can harness tools of technical power to its objectives (i.e. consistent quarter-on-quarter growth), but most people's ability to *adapt* can barely keep pace.

Yet, if digital technology reconfigures ambient human relativities faster than sensibilities can come to grips, the result may be a dangerous imbalance, a psychic gap filled by easily exploitable fears, the search for simple certainties, and far greater sociopolitical unrest. Carl Jung, the great Swiss psychoanalyst, observed in the early part of the twentieth century that:

> we have plunged down a cataract of progress which sweeps us on into the future with ever wilder violence the further it takes us from our roots ... It is precisely the loss of connection with the past, our uprootedness, which has given rise to the 'discontents' of civilization ... and to such a flurry and haste that we live more in the future and its chimerical promises of a golden age than in the present ... with which our whole evolutionary background has not yet caught up.[36]

It took the explosions of two world wars finally to resolve disjunctions set in train by the industrial age. What next? Marshall McLuhan, that metaphysician of the media, was one of the first to appreciate how much more profoundly the *electronic* revolution will disrupt our collective psyche. He characterized the late twentieth century as a period in which many people tend either to cling to shreds of outmoded identity or to hunt desperately for answers in new -isms and ideals. He was convinced that 'the agony of our age is the labor pain of rebirth.'[37] But the more fragmented and discontinuous life grows, and the more urgently people seek moorings of stability, the more volatile the situation becomes. Manifold humiliations of the human spirit are triggering explosions of violence on all sides. In the world of Islam (and even in 'the West') there are signs of 'a reasoned rejection of materialism and lack of clearly defined moral principles,'[38] and existing leaderships are regarded with a suspicion accorded to the ethically unhinged. Alienated citizens are turning away from politics in disgust; people are driven to extremes in despair; the globe is fractured by reactionary assertions of 'fundamental truth.' In short, the real chaos is human: it is merely being amplified by technical means.

Set against all of this is a new and seductive promise of escape via the digital frontier: a place of alluring but dangerous amnesia.*

Like a video game with all its distracting bright lights and magic, it beckons mankind toward a wonderland of fantastic promises that serve to distract from more immediate and troubling choices that concern terrestrial life. The Anglo-Dutch writer Ian Buruma once remarked that 'fantasy is the last recourse of the dispossessed.'[39] If so, then video-game fantasy is precisely what the

*The attraction of such dream worlds may lie at the very core of the primordial subconscious. Carl Jung well understood how different routes of psychological development could lead either to death or to rebirth. To him, the metaphors of classical antiquity held a key. He cites one tale in which Theseus and Pirithous follow a path into a deep chasm that leads to the underworld, where they hope to find and abduct Persephone. The effort proves exhausting. They seek a comfortable boulder on which to rest – to forget their trials for a moment and catch their breaths. But to delay the trials of transition, however briefly, is to risk the entire enterprise. Indeed, as the two travelers attempt to rise from their comfort and press ahead, they find that they have grown fast to the rock. The message is clear: for individual personalities or entire cultures, the process of self-transformation demands that we bravely confront the innermost

recesses of our cultural being. The depths both fascinate and horrify. 'Whenever some great work is to be accomplished, before which a man recoils, doubtful of his strength, his libido streams back to the fountainhead – and that is the dangerous moment when the issue hangs between annihilation and new life,' Jung writes. ' ... if the libido gets stuck in the wonderland of this inner world, then for the upper world man is nothing but a shadow ... but if the libido manages to tear itself loose and force its way up again, something like a miracle happens: the journey to the underworld was a plunge into the fountain of youth, and the libido, apparently dead, wakes to renewed fruitfulness' (C. G. Jung, *Symbols of Transformation*, trans. R. F. C. Hull (Princeton: Princeton/Bollingen Series XX, 2nd edn, 1967), p. 293.) The cybernetic 'revolution' is arguably forcing a similar confrontation upon mankind: a choice between accommodating terrestrial limits or turning to a forgetful escape that will only magnify the chaos instead.

world must try to avoid. The real issue on which tribal well-being now depends is not whether one all-conquering world-view is simply replaced by another. Even as the Cold War technocrats give way to the cyberdeck elite, the question is whether a more fundamental global balance can be restored between the imperatives of conquest and possession, on the one hand, and principles of nurture and harmony, on the other. Nowhere will this challenge be more pronounced than in the cut-and-thrust battlefields of electronic enterprise.

3
Castles and Abbeys

I am in the position of Louis Pasteur telling doctors that their greatest enemy was quite invisible, and quite unrecognized by them. Our conventional response to all media, namely that *it is how they are used* that counts, is the numb stance of a technological idiot. For the 'content' of a medium is like the juicy piece of meat carried by the burglar to distract the watchdog of the mind.

MARSHALL MCLUHAN[1]

Imagine yourself cycling down some sun-dappled back road, weaving between the fortified hill towns of southern France. Inspired by the landscape, you and your companion turn to talk of the Middle Ages. Your mind conjures a sense of disorder and fragmentation. You see images of castles, of mailed knights battling for the roadways, and of serfs working the lands you pass through. You think how, at that particular time, most folk were bound in feudal service to lords who, in their turn, owed fealty to kings. It was an age when the strongest imposed their rule; when the pre-eminence of church and monarchy formed a nexus of power that bounded the aspirations and the duties of the population, and that rigidly defined the material and moral context of life.

Every age has its own ruling hierarchy – traditionally an uneasy combination of those who dominate the scarcest material commodities (together with the means to move and protect them) and the priestly cults who guide spiritual life. In the Middle Ages everything hinged around land, and the notion that the king and his lords ruled by divine right (defined, of course, by Catholicism). Even today, cycling the back roads of France, I can glance over the landscape and easily pick out the physical remnants of the feudal condition: the fortresses, the monasteries, the fertile patchwork of land, and the roadways that brought goods to market. Later, industrialization replaced the landed aristocracies with new princes of corporate enterprise. Today the prevailing power landscape is marked by parliaments, crisscrossed by asphalt, and punctuated by the skyward thrust of office towers.

Or so it seems. But in fact the world is embarking on an extraordinary technical adventure – one that will erode even these familiar (and visible) monuments and replace them with structures of dematerialized electronic might. The much-trumpeted shift to an informational 'mode' of life is producing a new economic order with different and more abstract manifestations – not to mention its own quasi-religious orthodoxies. For example, Martin Bangemann, an influential member of the European Commission, states that 'the use of information will be *the* competitive element for industry for decades to come.'[2] Only a wholesale commitment to digital networks, or 'superhighways,' will assure twenty-first-century economic vitality. From Europe to Asia and throughout the Americas, this commitment has attained a sacrosanct and untouchable status: it has been elevated onto the altars of 'national security.' *Wired* magazine meanwhile suggests that the digital revolution is already causing changes 'so profound, their only parallel is probably the discovery of fire.'[3]

But if the world truly *is* embarking on an entirely new historical phase, what sort of hierarchies will it involve, and how will they take shape? What will be the cybernetic equivalent to the castles and abbeys, the roadways and pipelines that mark economic and political life? What forces, if any, are likely to dominate the landscape? What will this mean for citizens at large?[4]

The year is 1983, and your gaze is captured by an ad on TV. A young woman wearing angelic white running gear is breaking through a pair of steel doors into a packed assembly hall. She is hefting a potent-looking hammer as she tears down the aisle, her face set and determined. Shouldering past guards, she darts through an audience of gray-suited and vacant-eyed proles. Captive, they have been mesmerized by a message from Big Brother – a certain information monolith like IBM? – who is broadcasting his latest pronouncements over the giant telescreen. The young woman sprints toward the screen, coils into a spin, and hurls her hammer into his bespectacled face. This instantly splinters into a thousand shards, as does his grip on the crowd. The people rise. A deep voice intones, 'On January 24th, Apple Computer will introduce Macintosh. And you'll see why 1984 won't be like *1984*.'

The starting-point for advocates of the cybernetic 'revolution' is an incontestable proposition, perfectly conveyed in Apple's now-famous ad: digital technology and democracy are two sides of the very same coin. This is canon law – a defining precept that is accompanied by an odd, embarrassed silence if you happen to interrupt the litany and ask, How will this technology alter the relativities of commercial and political power? You are informed that this is a 'materialist's concern.' It seems the entire issue will somehow be technically invalidated. After all, adherents reason, didn't the personal computer blow away IBM's mainframe monopoly – and all its associated patterns of hierarchical centralization? This trendline can only continue. In a digital world, the individual will be progressively 'empowered.' Small will triumph over the large. *End of text.*

This cybernetic orthodoxy is marked by a naive faith in technically generated progress that would be touching if it were not so terribly dangerous. For one thing, it reflects an underlying conviction that the past has few (if any) relevant messages that apply to the situation today. In the isolated case when a historical analogy is brought into play, invariably you hear about Europe's Renaissance. This brought the invention of print and a consequent weakening of the Roman church's hold over textual truth. (Of course, the Renaissance also coincided with the rise of the Medici and other merchant princes, though this complicates the technical point and is therefore left unsaid.) When former Apple chairman John Sculley met with a group of photographers to discuss the ethics of digital imaging, some time ago, he redirected their attention to the fact that 'before Gutenberg's invention, less than 10 percent of the people were able to read, and those people were largely part of the Catholic church. Seventy years later, 80 percent of the population of Europe was able to read. We now have in the twentieth century another opportunity to pass along technology which had previously been in the hands of the few, just as reading was before the Renaissance: we can transform computers from data-processing engines and turn them into creativity tools.'[5] It would seem that, armed with these tools, anyone is a potential entrepreneur. The power of big organizations will no longer obtain. A few people – take the founders of Apple or Microsoft's Bill Gates – can change the world with a single inspired idea. The latest findings in chaos theory are cited as proof: how the mere motion of a butterfly's wing in Hong Kong can unleash a windstorm in Los Angeles. The process of change is 'unstoppable' and well under way.

Peter Schwartz is a cybernetic futurist who successfully predicted the demise of the Soviet Union while working as a strategic planner for that massive petrochemical giant Shell. He went on to become an independent entrepreneur at the head of the Global Business Network, and today enjoys an entree to the most rarefied heights of corporate and government power. He maintains that 'you can already see a definite trend toward smaller organizations.' He sees a seminal shake-up in the basic pattern of business organization – one that mirrors a deeper transformation in scientific outlook in which Newtonian physics is being eclipsed by cybernetics as the primary filter through which we interpret daily realities. 'The giant corporation, a symbol of the twentieth century, is now history,' he explains. 'The logic that led to its creation is no longer relevant. The old rules of the game were based on Newtonian principles – how long it took you to get from point a to point b, or how much energy was required to ship so many tons of x to place y. Now, more economic activity is informational in character, so those old rules no longer obtain. The logic of physics is irrelevant in a world where space and time no longer have meaning.'

And what about the bottom-line impact on power, one inquires?

'Power is being diffused down and out.'[6]

•

I recently sailed aboard a container ship out of Rotterdam harbor. An awe-inspiring juggernaut of towering hulls and high-loaded decks, displacing something like 300,000 deadweight tons, she seemed like an immense and yet virtually deserted floating metropolis. Now, in my mind's eye, she is both the perfect example of applied cybernetics and also a symbol of the more disturbing trends mainly left unaddressed by the emerging high-technological elite. This most modern of vessels is discharged and loaded automatically, by vast cranes, in precise and preplanned configurations. She drifts out of her berth in an eerie, inanimate quiet. The harbor pilot is soon dropped as the ship takes to sea. Her decks are deserted and the companionways are lit to an intense fluorescent pitch. This only intensifies the unsettling overall sense of emptiness. Then, standing with a pair of lone mariners on the bridge, I note the curious absence of traditional nautical fittings like the ship's wheel. The noon watch is bent over glowing computer consoles, typing in digital commands for transmission throughout the ship's labyrinthine network of automated servos that close valves, steer the rudder a few points to port or starboard, or raise the engine speed to 'full ahead.' When you look up, it is with shock that you find yourself on the high seas, so lulling is the hypnotic world of the screen.

Like this massive, exquisitely automated, but virtually lifeless ship, the cybernetic economy embodies a potential to be more productive than any comparable system known to man. But when you sift the fact from the hyperbole there is no detectable trend away from corporate bigness, as Schwartz and his fellows maintain. True, size is no longer the only factor in success. You can profitably build cars in smaller runs; specialty steels can be efficiently poured in 'mini mills.' But take a closer look at the automotive industry and you notice that even companies traditionally known for limited, quality output – companies like BMW or Mercedes – are *widening* their productive base. Ditto for consumer electronics, communication, Hollywood entertainment, pharmaceuticals, and more: the spiraling cost of developing new products and marketing them globally creates a powerful incentive to grasp economies of scale and scope. Nowhere is this more true than in the production of computer chips. For example, since the early 1970s the price of a megabyte of computer memory has plunged: from $550,000 to roughly $38 today. But the cost of building a factory to make these chips has sky-rocketed: from less than $4 million to over $1.2 billion now. With price tags like these, only the biggest can play.[7] The same pattern obtains through industry at large – *especially* the information industry. The trend is toward a simultaneous digitization of markets and a concentration of control.* In theory, these simultaneous trends make for an organization that is more 'adaptive' to the lightning pace of change. In practice, such speed and agility carry a high price. The faster and more maneuverable the aircraft, the more inherently unstable its flight (which fighter pilots refer to as 'managed crash').

*'Mergers and acquisitions reached record levels in the worldwide information technology business last year [1995] as companies sought the size and technology to compete in global markets,' writes Alan Cane in 'IT mergers reach record levels,' *Financial Times*, 1 February 1996.

The truth is that what is so fashionably mistaken for 'decentralization' is actually nothing more than an atomization of the modern corporation into functional parts that revolve around a common central processor (like a computer CPU). What we now refer to as 'an airline,' for instance, will be broken into a component leasing the necessary aircraft, another responsible for managing the fleet, a third for handling maintenance, another still for pilot training and crew management, and so forth. Each of these components now has to pay its own financial way. Should it falter, it cannot assume help from other nodes operating at their peak. At the extreme, in fact, there is always a chance that the corporate hub will reroute its resources elsewhere. The organizing core can arbitrage between sources of labor, finance, and components on a global scale. (Of course, it may not find it necessary or expedient to do so. However, the knowledge that it *could* tends to exert a chastening discipline on the nodes that depend upon it.†) They must all become every bit as versatile as the movement of data is quick: the cybernetically wired organization is vitally predicated on efficient data control. Says management guru Peter Drucker, corporate re-engineering

†There is some question as to how much new 'globalization' has actually occurred: evidence suggests it is mainly confined to finance and is otherwise limited in the case of industry at large. One recent study points out that about three-quarters of the value ↪

added by multinational corporations is generated within their own home markets (Paul Hirst and Grahame Thompson, *Globalization in Question* (London: Blackwell, 1996)). Even the largest networked enterprises depend crucially on close ties with home government, stable industrial relations, and dependable sources of physical supplies. The truth of the globalization thesis is that it offers big companies an important bargaining chip in their negotiations with suppliers, organized labor, and governments, according to Winifred Ruigrok and Rob van Tolder ('The myth of the "global" corporation,' *The Logic of International Restructuring* (London: Routledge, 1996), pp. 152–68). It also makes it easier for politicians to evade responsibility for deregulatory acts on grounds of 'defending global competitiveness.' This is ironic when you consider how extensively the financial and corporate spheres have made use of tax-payer-funded public largess. 'At least twenty companies in the 1993 *Fortune* top 100 would not have survived as independent companies if they had not been saved in some way by their governments' (ibid., p. 217).

is now all about 'changing an organization from the flow of things to the flow of information.'[8]

Once an organization has been wired and reconfigured for efficiency, there will be fewer human tasks to go around. 'People are now becoming the most expensive *optional* component of the productive process and technology is becoming the cheapest,' asserts Michael Dunkerley, a software specialist who recently published *The Jobless Economy?* 'People are now specifically targeted for replacement just as soon as the relevant technology is developed that can [replace them].'[9] In the view of management guru Charles Handy, top executives will become quasi-partners in the networked enterprise. 'Organizations could ultimately become a collection of project teams, harnessing the intellectual assets around a task or an assignment, rather as a consultancy company or an advertising agency does now ... project *leadership* will become the key to corporate performance.' Handy foresees a form of corporate federalism 'built on shared power, compromise and negotiation.'[10] For those who survive the cull, it will indeed seem as if life has been opened out and democratized. But for those layers of middle management that have been cut out, and for what remains of the old 'blue-collar' workforce, whose bargaining power has been radically curtailed, it will be a different picture indeed.

With a tumultuous structural transition sweeping through industry, job responsibilities and production processes are being relayered according to a new (computational) design. But there is little evidence to suggest that power is devolving 'down and out' to the average employee. Take a close look at the Swiss-based engineering company Asea Brown Boveri (ABB). It has been fragmented into countless subunits or profit centers that produce industrial robots, power generators, railway engines, and the many other wonders of modern electrical design. The group generates yearly sales of $34 billion and boasts a far-flung worldwide work-force of more than 210,000 souls. Aspects of operational manage-ment have been thrust further out toward the front line. (This is an example of 'empowerment.') Peers can communicate with peers without going through a hierarchical sieve. (As one member of the digerati quips, 'E-mail flows through an organizational chart like meat tenderizer.'[11]) Yet the entire enterprise is orchestrated by a group of executive coordinators working out of the Zurich HQ. Like mariners on the bridge, the essential task of this celebrated team is to synthesize data from the various corporate subsystems and fine-tune the ship's course under the overall direction of their mustachioed chairman Percy Barnevik and his Swedish chief mate, president Göran Lindahl.* ABB's command and control structure may now appear more flat than pyramidal, since whole layers of middle management have been strip-ped away. But, even as front-line managers assume more responsi-bilities and risks, at the end of the day it is only ABB's top officers who answer to the shareholders, just as a ship's captain is ultimately responsible to proprietors in port.† The enduring need for clear lines of responsibility and control in business, both legal and fiduciary, means that real authority is still very much vested in the core.

*In fact, such business systems have close parallels in the world of role-playing games and multi-user dungeons (MUDs), where powers of control usually reside with the 'wizards' and 'gods' who condition the range of possible player interactions.

†Harvard professor Robert Reich has famously described such executives 'symbolic analysts' and 'intellectual mercenaries.'

Seen from this perspective, a cybernetically wired company dis-plays attributes remarkably similar to those of the computer systems that drive it. The flow of materials and products – like data – can change with bewildering speed. Automated 'slave processors' actually carry out the ever-shifting range of commands. The whole affair follows guidelines – like software routines – designed and orchestrated in the executive suite. This is where the system architecture, the processing priorities, and the pattern of data flow are set. And of course the overall integrity of the system crucially requires that resources be routed with great haste from any 'nodes' – i.e. subsidiaries and people – that underperform or fail.[12]

So: in the Middle Ages it was the control of land that conferred power. During the industrial era it was dominance over labor and material inputs. In an informationally configured economy, in contrast, authority belongs to those who command synthetic skills: the ability to access key data and *manage it* in pursuit of their aims. Since so much more raw information will be produced, *relevant* facts will attain a much higher relative scarcity value. As Apple's founder Steve Jobs notes, 'information's usually impossible to get, at least in the right place, at the right time.'[13] Thus, the more complex and geographically diffuse a networked corporation becomes, the greater the influence of those with management oversight over the whole. Likewise, those who remain submerged below the waterline are unable to anticipate and act on climatic shifts. Information *about* information is becoming essential – a capacity to see over the waves and make out the changing shapes of weather and the horizons ahead. As Cyrus Freidheim, vice-chairman of the international consulting firm Booz-Allen & Hamilton, remarks, 'The fundamental decision about the allocation of resources always remains centralized, no matter how many decisions are pushed "out into the field."'[14]

I have to confess to a passion for vintage prop planes. A special favorite is the DC3, a craft of personality and a gorgeous simplicity of design. The plane flew onto the scene in the 1930s. The fledgling Douglas aircraft company had captured a dream order, from a then equally freshfaced airline called TWA, to develop an entirely new kind of machine. The DC3 was the result. It was christened the Dakota by British paratroopers in the Second World War. An elegant, twin-engined craft, it is now the object of cult-like admiration from both aircraft aficionados and Latin drug barons alike (though for different reasons, needless to say). To me, the plane captures all the exhilaration that surrounded the pioneering days of flight, and I also have a visceral love for the

DC3's curves, and the noble way it glides through the air, striking a clear-headed pace and sensible altitude that agreeably contrast with the thrust of today's frenzied, hurtling silver tubes. Pilots, on their part, appreciate the plane's robust construction – almost 4,000 are still in regular service – and its versatility in difficult terrain where rugged landing-strips are the norm. They say it is a craft you can fix, without high-tech whiz-kids, when the odd thing goes wrong now and then.

The commercial importance of the Dakota is of another order, though: with its (then) radical contours, high passenger capacity, and low fuel consumption, the Dakota revolutionized the business of air travel. Traveling by air is something we take for granted these days, but before the DC3 it was a costly and dangerous affair. Passengers were layered in heavy blankets against the cold. They were expected to endure noisy cabins and otherwise rudimentary facilities. Stuffed cheek to jowl with packing-crates, these intrepid travelers represented nothing more than an adventurous diversion for airline managers, who at that time drew the lion's share of their revenue from government-subsidized air mail and freight. Then, suddenly, the DC3 made passenger flight a paying proposition. The DC3 not only inaugurated commercial aviation, it also foreshadowed the demise of the great transatlantic ocean liners and the many ancillary enterprises grouped around them. Quite unexpectedly, a new technology had shifted the dominant *context* of the market, moving it from sea to air. This, in turn, opened a virgin frontier of opportunity for those with the right skills and tools.

Now, unlike the DC3, with its powerful impact on one industry (travel), cybernetic technology will reshape the entire corporate landscape. Indeed, from the start, it has exhibited this Houdini-like knack: once insinuated into a business environment to automate some narrowly defined task, like the corporate accounts or word processing, the cybernetic tool winds up interacting with other systems, hooking up with other machines in networks, transforming the traditional environment and eventually defining a new one altogether. Of course, every business still needs labor, raw materials, and other physical inputs; indeed these can be decisive to ultimate success or failure. Yet, even if it continues producing the same things (if in different ways), a company's viability and competitive edge will lie in controlling *binary* streams. This is why philosophers of cybernetics – visionaries such as Ithiel de Sola Pool – insist that 'information is everything.' It is like a universal and supremely malleable piece of clay: almost any 'input' can be represented by an informational equivalent, from a butterfly wing to a nuclear detonation, and be expressed as a binary string of zeros and ones.* Cybernetic tools can just as easily control an airliner as guide a ship; they can display air currents acting on a wing or the temperature inside a boiler; and they can animate any automated mechanism, from wing flaps to flues. By simulating hypothetical situations, saving time and resources, a printing-house can fiddle with different layouts before committing to print. A car manufacturer can switch from one component supplier to the next as prices and currency values change. The entire *process*, the countless complex procedures that make up the unity of any enterprise, can be rationalized with astonishing effect.

'It's like DNA and gene technology,' an executive enthuses. 'Once your enterprise is wired, there's nothing to stop you from recombining its constituent parts in wholly new ways.'

And so we're confronted with this great and curious paradox: as the functional environment of business is progressively miniaturized, crammed onto microchips, and compressed into opaque lines of code, the scope and resonance of the technology are exponentially widening to the point where, having started in discrete situations, the technology is now ubiquitous and all-powerful. Herein lies the single most unsettling trend: as computers infiltrate themselves ever more thoroughly through the anatomy of

*Of course, a digital representation of anything – voice, picture, DNA sequence – is meaningless if not wrapped in explanatory software, which thus acts as a medium, or a vessel of conveyance, with certain unique qualities of its own. As the reader will soon see, moreover, what crucially differentiates the processing of binary simulations from the fluttering of real-life butterfly wings is that the former are bound not by any natural restraints but rather by rules implicit in the way they have been coded by software programmers. Thus, the process of representing the world in informational terms, of translating the natural into the binary for the sole purpose of control, also subtly reinforces a dysfunctional tendency to view nature and all its life forms as a sort of raw material that may be shaped and manipulated at will, rather than as an inheritance that is best handled with reverence and care.

enterprise, and from there spread throughout the economy and into society at large, they start to form the electronic nervous system of modern life – a nervous system thirsty for the animating life-force of binary code. Just as companies are becoming little more than computers, *national economies* are also being absorbed into a vast and globally networked grid. What values will characterize these elusive new circuits of power?

On my cycling tour with Miss B. in southern France, we came upon a small village one brilliant Saturday morning and found ourselves absorbed in the modest pageant of its market day. There were flowers in the sun; stalls of vegetables, fruits, sausage, honey, and herbs. Here, in the heart of the Haute Provence, rural folk were gathered in lively groups at the cafés and around the square. No one seemed to be in any headlong rush to transact their affairs. Or perhaps this was the essence of their affairs: exchanging opinions and local gossip, remarking on the state of the weather and crop subsidies, were all inseparable from the business of *life*. The whole scene was seemingly disorganized, scattershot, no doubt incredibly inefficient (from the industrial cost-benefit point of view), but it suggested a refreshing and alternative way. In fact this was precisely what attracted us in our search for a short break from the sprawling pressures of urban life.

Ruminating over what we'd seen and heard – gazing all the while at the startling and archaic apparition of a man who was actually watering a horse at the village fountain – Miss B. and I fell into a conversation about life and the land. This interdependence has defined most of human experience. In this pocket of France at least, daily existence, subject as ever to the caprice of distant powers, still follows a pace conditioned mainly by the procession of seasons. In a fiercely wind-torn and changeable climate, the environment is indomitable: life-giving and death-dealing in ever-fluctuating degrees. The novelist John Berger, who lives in a different part of rural France, describes its 'culture of survival,' where the procession toward an uncertain future involves a sequence of repeated acts of enterprise aimed at pushing a metaphorical thread through the eye of a needle. Harsh ethics are drawn from nature: poised against the remorseless cycle of change are one constant – honest work – and a richness of meanings derived from ritual and evolved routines.[15]

Most of us who live in cities are more familiar with the imprint of modern industrial civilization, with its largely predictable schedules and standardizations that gradually emerged along with mass production. Our time is broken down into precisely measured segments. Life is built on the assumption of continual economic expansion, which in turn depends on our continued pursuit of individual material gain. None of this happened overnight, of course, but rather in spurts which built to a cumulative effect. Today, for most people, the affairs of business infrequently coincide with the business of one's personal life.

Relaxing in a French café, sipping a *grand crème* and reflecting on the contrast in lifestyles, I couldn't help but speculate how an even more headlong rush into an information economy, compressed into decades rather than spread over centuries, will actually re-configure the current patterns of human experience. If the developed world becomes more informational and less oriented toward consuming resources this will only reduce our burden on the environment. Or so it seems – for the single most significant application for cybernetic systems (apart from personal entertainment) is actually to 'rationalize' the industrial process and make it cheaper to produce *more* goods for sale. (The volume of seaborne trade, for example, climbed by over a third, from 3.09 billion tonnes in 1983 to 4.47 billion tonnes in 1994, with the fastest growth coming in manufactured goods. This figure is expected to climb to nearly 5 billion tonnes by the decade's end.[16])

Leaving aside the underlying question about whether this consumptive path is the best way forward – something every person decides on their own – it is almost certain that patterns of human livelihood are going to change as profoundly as the economic structures on which they are based. The mechanics of the production line – first elaborated by American firearm manufacturer Eli Whitney, refined in Chicago slaughterhouses, and finally perfected by the auto-maker Henry Ford – have left an unmistakable imprint on contemporary experience. Similarly, wired enterprise, and society at large, will be conditioned by the values inherent in the cybernetic media that drive it. These include:

mobility and abstraction
acceleration
polar fragmentation

Mobility and Abstraction

Let me take a short digression with a story about Roberto Calasso, an Italian essayist and publisher who is a connoisseur of great books in their original editions. 'There's something very important even in just living with certain books,' he maintains, describing 'the sense of familiarity with these things that you can't get when they're behind glass ... When they are part of your daily life, when you take them in your hand and open and shut them, you understand something even before reading them.'[17]

Among other things, Calasso is concerned with the subtle tie between *content* and *context*, a vital unity that is easily severed in cybernetic environments. You can read a book in any cover, and you usually pour wine from a bottle, not from the original cask. But no matter how far they travel or how variously they resonate, neither the book nor the wine can ever be separated from the place from which it came. *War and Peace*, in Tolstoy's finished form, cannot be made more 'interactive' than it already is: change the ending and you produce a different work. A fine wine can be enjoyed anywhere, but it is no less French for its being savored by the nominally communist rulers who occupy 'people's palaces' in the Forbidden City of Beijing.

Now consider digitized library cards. Even if all of the data are transcribed without error (and the error rate is certainly high), the new record cannot capture the many essential signals of the original – everything from marginal notations to smudge marks that form a treasure trove of contextual information for scholars doing research. Artificial-intelligence tools can correct filing mistakes (whether it's the variant 'Tchaikowsky' or 'Tsjaikovsky' depends on time and place); but, on the other hand, they create new muddles as well – like globally modifying *all* Madonna references to 'Mary, Blessed Virgin, Saint.' Whoops![18]

When it comes to cybernetic business systems and their cultural cradle, the link between content and context is crucial. The bit-stream that flows through a computer is composed of indifferent zeros and ones, universal symbols whose essence is that they can be economically exchanged with other computers, near or far. In the process of being captured in code and shunted about in these ways, the content of a given message becomes little more than a displaced husk, a rootless abstraction stripped of defining spirit. Take the simple sentence 'Look before you leap.' It's charged with entirely different overtones depending on whether you're chatting with a commodities broker, a bungee jumper, or a convicted felon. A rebellious graffito scrawled in the humid heat of a *barrio* night in Rio will be pregnant with prophetic contextual meanings; these are lost when they are distanced as mere data on-screen. Likewise, the languages shared by Englishmen and Americans use identical words to imply different things. These pivotal distinctions are scattered in the placeless world of cybernetic enterprise.

Now consider the procedures of a given business – say food retailing. Americans have a unique way of distributing groceries to supermarket shelves. It is highly elaborate, and it crucially depends on the existence of *a supermarket style of life*. Now, as long as Japanese people persist in buying their groceries from small, family-owned stores, the American distribution system will never work in Japan. However, if you can transform an economy in such a way that its businesses must operate according to universal and *computational* standards, then slowly and gradually the local context can be reconditioned into a more cybernetic shape.* When the French hypermarket, being more 'efficient' and thus cheaper than village shops, displaces the old retail network, it also helps to unravel the entire

*US and EU trade delegations visiting Tokyo in 1994 exerted strong pressure on the Japanese government to 'end the mass of planning and operating restrictions that impede the growth of supermarkets, a move which would spell death for the 1.5m small, family-run businesses that dominate Japanese high streets, but high profits for the budding discount sector' (William Dawkins and Michiyo Nakamoto, 'Japan under pressure on two fronts,' *Financial Times*, 15 November 1994).

fabric of life built around the town square, where people gather not only to buy their groceries but also to socialize, to play a game of *boules*, and to sip coffee and watch the world go by.★ Such pastimes are rendered 'obsolete.' As one Nissan executive, busy building a car factory in Mexico, hurriedly explains, 'We do not care about the Mexican way of doing things. Anyone who works here has to think only about international standards ... otherwise there is no way we can compete.'[19]

★I've deliberately emphasized this word 'efficient.' What does it mean? According to what timeline (and what baseline) is it measured? Clearly, computationally measured efficiencies reduce immediate corporate cost and enhance short-term growth; but they will also sometimes have highly inefficient long-term cultural and ecological consequences. Other activities – like the sacrifice that underpins all successful communication between individuals and between cultures – can only be termed 'inefficient' if considered from a binary point of view: that is to say, if broken out of the holistic context of life. This particular approach allows decisions to be taken without regard to the full range of trade-offs and costs implied.

This place-insensitivity, this process of importing and exporting of business 'systems' (and all their subtle but unacknowledged biases) without regard for their impact on the human context in which they are 'installed,' can only accelerate as companies are transformed into vast data-processing engines fueled by binary universals. Already, almost by default, wired transnational enterprise is projecting imperatives that assume an unimaginable degree of market conformity: in the global supermarket, defining human particulars can be hosed down and drained through the grate. With its incomplete acknowledgment of unquantifiable local realities, it is like a distant suitor trying to consummate a relationship over the phone: no matter how 'connected' he is, the myriad richnesses of sense and sight that accompany any physical encounter are simply lost over the line. Indeed, the most worrisome trend is that, having translated the particulars of business into numerically defined abstractions, companies hacking their way through competitive digital combat too easily forget what differentiates binary patterns from the human realities they convey. Thus distanced, they feel less compunction about acting upon those abstractions – for instance by 'deactivating a functional business component' employing 12,000 – in irresponsible ways. This is how human communities are uprooted and plunged into a tumult of digital discontinuity.

Acceleration

Some years back, in a small town outside of Bologna in Italy, I was taken on a 'little run' by the maniacal test driver at Lamborghini, the manufacturer of hand-built and astronomically expensive racecars like the Countach, an angular machine boasting twelve cylinders, 455 horsepower, and a profile more akin to a tactical jet fighter than an automobile. As the driver racked up through the gears, the machine began to emit a sort of nervous whine that rose up the register with the demented insistence of a giant teakettle coming to boil. Suddenly we were power-sliding around a broad curve at something like 180 miles an hour. Eyes wide, fingernails dug deep in the supple leather of the passenger seat, my senses fragmented by jagged streaks of passing country, and my guts wobbling out of control, I decided that, whatever else happened, this had to be the ultimate experience in speed.

I was wrong.

In a car, the engine can spin low revolutions in top gear. The computer has a straight 1:1 ratio – so, the quicker its clock speed, the faster it rolls. Now, processing power has been doubling every year for the last decade, according to the dictates of Moore's Law.★

★'In 1964, six years after the integrated circuit was invented, Gordon Moore observed that the number of transistors that semiconductor makers could put on a chip was doubling every year. Moore, who co-founded Intel Corporation in 1968 and is now an industry sage, correctly predicted that this pace would continue into at least the near future. The phenomenon became known as Moore's Law, and it has far reaching implications.' (G. Dan Hutcheson and Jerry D. Hutcheson, 'Technology and economics in the semiconductor industry,' *Scientific American*, January 1996.)

Since every click in the computer's clock is like one pace for a marching troop, there is now a vast army of bits and bytes skipping along at around 40 billion drumbeats per second. If these magnitudes seem inconceivable, consider the effect: the faster the machine, the faster the potential pace of the enterprise it drives. Products can be developed faster for shorter lifecycles and tighter markets: *The Maradona Cookbook* appears the same week as the football World Cup. Trying to keep pace with the latest developments in PC technology is like chasing the tail of a rocket: these machines are obsolete before they're shipped out the factory door. They are assembled to ever-shifting specifications with essential components delivered 'just in time' from around the world in a trend called 'mass customization.'

As the cycles of global competition accelerate, ever more capable systems are wheeled into action to overcome the challenges caused by the ones that came before. Suppose you're a telephone company and your competitor starts offering cheap long-distance rates. You set your propeller-heads to work writing network software, and with luck and a couple of weeks you'll be ready to offer customers an even finer deal – say, a 20 percent discount on the five numbers they phone most. But then your competitor crunches more numbers and comes slugging back with a modified scheme. Back and forth it swings. The scale and velocity of processing power that can be brought to bear on any business problem mean that the intervals of competitive advantage are growing shorter all the time.*

It is on the human scale that the costs of this acceleration are most pronounced. I live across the street from a fairly sizable financial conglomerate, and the traders who stumble out its doors at the end of every day, weighed down by their briefcases of homework, look as if they've been agitated in a vat of concentrated bleach. Like other front-line managers, pale-faced and determined, they are now 'empowered' to take instant decisions that optimize efficiency and augment corporate gain. Similarly, the store manager at my supermarket now uses a database to fine-tune the product mix on his shelves to what they call 'real-time microshifts in consumer demand.' But all of these people are working harder just to stay in place. In a world of business 'benchmarking,' the benchmark of success seems to be set higher with every incremental advance in cybernetically generated productivity. So, as the flywheels spin ahead, ever faster, the urgencies also multiply: develop fresh products, deliver to market, move on to the next promising frontier. The human toll extracted by these ever-spiraling expectations of

*This compressed interval between business rise and fall is often cited as an opportunity for outsiders and innovators. Examples include the much trumpeted ascent of Apple following its (short-lived) defeat of IBM's mainframe monopoly, the proverbial victory by Wal-Mart over its archrival retailer Sears, and, of course, Microsoft's meteoric rise to software ascendancy from a standing start. But these merely typify the early phase of any industrial 'paradigm shift.' As it evolves, the fastest, the most mobile, and the financially most powerful enterprises will pull ahead (an issue that is discussed at length in Chapter 4).

achievement – for example, less time spent with family and friends; drug abuse; violent crime – arguably constitutes a new form of enslavement. As the figures mount on the corporate bottom line, society is saddled with dangerously escalating levels of stress and insecurity.[20] How can we strike a better balance between the desire for growth on the one hand and broader priorities such as social equilibrium and ecological conservation on the other? Will we even be inclined to search for answers when so much of life is taken up with the struggle to meet the accelerating short-term objectives imposed by our own machines?

Polar Fragmentation

One of the most remarkable trends in advanced computer research is a repudiation of attempts to *engineer* software programs in favor of *evolving* them through 'biological' competition. Step by step, one line of code is tested against the next: the fittest survive, and the most robust come out on top. This Darwinian techno-selection has already been ardently embraced by industrial enterprise. It is being used by pharmaceutical groups to model and produce powerful new drugs through electronic simulation. What gives the resulting product its wallop is often shrouded in mystery, since the un-controlled development process is so unimaginably complex. But the chemical companies are unconcerned. The point is: *the stuff works*. And it sells.

Kevin Kelly, editor of *Wired*, explores some of these evolving intricacies in his book called *Out of Control*.[21] For him, the meta-morphosis of commercial enterprise from an old industrial model to the new computational mode, fed by abstract coded sequences racing like a torrent among countless competing circuits and nodes, is yielding biological complexities too immense for comprehension or management:

> Think of our own mind. You probably go around saying 'I' do this or that. But there's no 'I' in your brain. It's just divided up into a lot of little tiny minds, some of which are so dumb they don't even

think. The only way you get intelligence is to have a society of little minds. Same with this network of many minds and computers we're connecting together around the world. It will be a supermind. It will produce thoughts on a level no human will be able to understand or even perceive, and the emergent phenomena will be out of anyone's control.[22]

But these protests of helplessness fail to convince. For one thing, such a chaotic mode of discontinuous change, this modern-day version of 'divide and conquer,' is anything *but* inevitable. Rather, it is the product of *choice*: it is the outcome of a singular pattern of economic development that depends upon ever-expanding rates of turnover and growth. In order for this to gather pace, cultural and spiritual impediments must be stripped away.

Ironically, it is precisely the systems of human society and natural ecology that are most resistant to control. The computing environment, for all its complexity, is also crucially one of *programmable* activity. Say a major car manufacturer decides to launch a new car. Design work might be farmed out to one of several independent contractors based in Turin. Testing and simulation contracts are awarded to a company in Seattle. Production is played off among various competing assembly sites, depending on variable factors like currency exchange rates and labor costs. There will be 'just-in-time' supply of countless parts from a network of subordinate producers. Now, all of this may be geographically distributed and complex, to be sure. But the energy that holds it together – and thus its real motive power – now resides in a network and in software routines that have been deliberately configured to achieve the specified objective. If these systems resulted in chaos, executives would be sacked and/or the company would fail. Instead, we can look around and see that networked industries such as aircraft production, petrochemicals, retailing, and computing are now consolidating and growing in scale.

At the same time, however, the organizational structure of industry is being fragmented and workforces are becoming more polarized. Intel, the leading computer chipmaker, predicts that the wiring of commerce will sweep away the middlemen and distributors who today help deliver products to market and deliver signals about price and demand back to the manufacturer. These roles will be performed instead by efficient software and systems. Economies will continue to grow. There will be more productivity 'per head.' Large corporate systems will become stronger and their host nations therefore supposedly more 'competitive' on a global scale. Yet this revolution triggered by corporate networking will require the fragmentation of mutual support commitments, and the unequal relayering of risks and opportunities both within organizations and in society at large. Empowerment, for those at the top, will be derived from their abdication of responsibility for dependents further down the scale. Growth, in short, will become vitally dependent upon fragmentation and polarity.

As David Korten writes, we are evolving:

> a dualistic employment system. The employees engaged in the core corporate headquarters are well compensated, with full benefits and attractive working conditions. The peripheral functions – farmed out either to subordinate units within the corporation or to outside suppliers dependent on the firm's business – are performed by low-paid, often temporary or part-time 'contingent' employees who receive few or no benefits and to whom the corporation has no commitment.[23]

Thus, a golden interval of history in which a portion of mankind has enjoyed a relatively equitable distribution of wealth and opportunity is now giving way to an 'upstairs/downstairs' future, a tyranny of monotonous, quantitatively defined systems sweeping over the qualitative diversities that enrich our kind. The nervous gap between the metaphorical 'zeros' and 'ones,' the winners and losers, may lead a cybernetically driven economy to hang or crash, like the computer it has, quite literally, become.★

★The filmmaker Woody Allen, in one of his great bittersweet classics, plays an inveterate New Yorker who admits, 'I'm sure there's a great beyond, but I want to know just one thing: how far is it from Midtown and does it stay open late?' In a wired economy, the great

What is missing and must still be introduced into this web of competition and global interconnection is an acknowledgment of mutual fate, of shared sacrifice and gain, of the contextual obligations that adhere beyond is where the downsizees reside. Money, skilled knowledge workers, and flexible production 'nodes' of a wired economy are now maneuvering with more freedom than ever before. A great deal of the mid-level management is being cleared from the scene, its jobs replaced by machines. What obtains is a sharper dichotomy between the upper echelons (with loyalties tied less to community than to corporation) and their redundantly skilled and immobile compatriots. They have little contact: each exists for the next in a proverbial 'great beyond.' For example, between 1975 and 1995 the output of the UK economy climbed by 45 percent. But the number of people with work advanced by just 2 percent. The 'efficiencies' delivered by informational networking have thus failed those whom it is supposed to benefit. True, those with jobs have done relatively well. Between 1978 and 1992 senior executives' pay rose 50 percent; but the pay of the lower echelons remained unchanged. Meanwhile, in America, the world's most prosperous economy, there are now estimated to be upwards of 60 million poor. Of these, 26 million – half of them children – depend on soup kitchens and feeding programs to survive.

to opportunity. Only with these crucial but incomputable values – conveyed only through human communication – can evolutionary chaos be channeled toward civil stability.

See Peter Davis, *If You Came This Way: Journey through the Lives of the Underclass* (Chichester: John Wiley & Sons, 1995). Pamela Meadows, director of the Policy Studies Institute, a UK think tank, warns that if prosperity is not more equally shared 'we may eventually end up paying the costs of social dislocation in higher welfare bills, rising levels of crime and other forms of social malaise' ('When growth fails the unemployed,' *Financial Times*, 27 February 1996).

So, it now seems unlikely that cybernetic technology will relegate large organizations to the scrapheap of history: it will simply disperse them into a constellation of less visibly related nodes. Likewise, there is little evidence that individuals operating within such organizations are being positioned to decide their own working conditions and destinies. On the contrary. While the Old Guard is being replaced by a generation of Young Turks (who happen to wield more advanced and far-reaching tools), the leverage of most employees has declined even as their treadmills are 'upgraded' (or speeded up) by technical means. The magical agencies of cybernetic technology are reconfiguring, not eliminating, the relativities of power. The essential matrix in which business, politics, and society have their existence is subtly migrating away from visible intersections in a physical world built to human scale – one where the ancient fortress and the contemporary pipeline each has its story to tell – and into a dematerialized computational realm. Roadways of cybernetic commerce are being etched into microscopic traceries of silicon. Contemporary castles and armories are being built on a foundation of circuits and complex programming routines. And a new elite is emerging – high priests and warriors rolled into one.

These are The Masters of Code.

4
The New Feudalism

The old greed for power, long ingrained in mankind, came to full growth and broke bounds ...

CORNELIUS TACITUS, *THE HISTORIES*[1]

Right up to its restructuring in 1995, American Telephone & Telegraph (AT&T) stood as a symbol of corporatism writ on the largest and most epic scale. Known as a conservative firm, steeped in classical traditions, it plumbed the myths of antiquity to find a symbol that would do justice to the glory of its self-image. Eventually it settled on Mercury — a bearer of messages — that Roman master of commerce and cunning. Today, a towering statue of this god dominates the headquarters entryway, sheathed in a burnished crust of 23-carat gold. His eyes look skyward and his arm is outstretched with a cluster of triumphal thunderbolts. Somehow, though, this gilded apparition seems misplaced amid the nesting birds and the landscaped greens of Basking Ridge.

I have come to these mist-shrouded New Jersey hills to conduct a battery of interviews. And, apart from this surreal clump of gold, the headquarters complex otherwise seems almost monastic and discreet. Here and there a flash of yellowed brick or some terra-cotta rooftiles peak out amid the grove of calm. Even as I'm primly ushered into the carpeted sanctum of the executive suite, I have a deceptive but lingering sense of joining a meditative retreat, of entering a floating temple of aesthetic or spiritual pursuit, not

the nerve center of a great corporate giant, a colossus of communications whose annual revenue, if still consolidated, would easily
overshadow the combined economic output of many sovereign
states.*

*In 1978 AT&T became the first corporation in history whose total assets exceeded $100 billion. AT&T reported sales of $79.6 billion in 1995, when it was split into three operating companies. Its cumulative turnover since 1984 is close to $800 billion. Its immense assets of technological know-how, which drive the business, have never been adequately quantified.

Having bought out or decimated all of its smaller rivals in the early part of the twentieth century, the 'Bell System' took its place among the great industrial-age mono-liths: built on standardization and efficiency. As with Henry Ford's Model-T (as the old joke goes), you could have any color telephone you wanted – as long as your choice was black. Standards were set in conclave. The network was the holy of holies. Ma Bell was the great mother church in an expanding (temporal) realm of telephony. Like every church, it was predicated on a body of cardinal beliefs. The most crucial was even cast into a broad bronze disc and originally set into the floor of its lower-Manhattan headquarters. Today, having been transferred to Basking Ridge, and worn smooth and bright with the passage of countless scuffing shoes, this oversized coin is a historical relic in every way. It reads, 'One Policy, One System, One Universal Service.'†

†This phrase was coined by AT&T's patriarchal chairman Theodore Vale. The company's status as a publicly regulated private utility, a franchise monopoly offering reasonably priced and universal service, was affirmed in the Graham-Willis Act of the US Congress in 1921.

This rings positively bizarre in our new world of triumphant individualism – a time when computers promise unique experiences for all. Mass *anything* is Orwellian heresy – the Oldthink of digital times. But pause. Consider. There is a public-service obligation implied in this dusty old phrase. The telephone net didn't grow smoothly, nor by magic, nor by technical default. It was the result of a bloody battle. First Ma Bell devoured her fledgling competitors; she became immense, bloated, obscene. Her executive bigwigs were forced to Washington, brought to a table where they had to negotiate a deal. AT&T would be allowed to keep its monopoly, but the network had to be universalized and its clients provided with service on equitable terms – both the marginal ranch in Montana *and* the wealthy customer with busy lines in corporate New York. Very similar guidelines prevailed in Europe and Asia, where the phone companies were state-owned, outright. This is why today's global network is still so strong – it is a united federation that forms the biggest and most elaborate machine ever built. It stands as the quintessential outcome of a cooperative *sharing* of cost and benefit. Nowadays, in the developed world where this co-operation prevailed, you can reach a phone, lift a handset, and ring whomever you please, regardless of their location, and whenever the spirit happens to move you. Meanwhile, four-fifths of the world's people still cannot afford to install a telephone handset, much less 'buy into the future' with a $1,000 PC.★

'*Deregulation, liberalization, and privatization*': the 1980s introduced a seductive new mantra into the realm of public affairs. First the feral energies of finance were released from their national bounds; now, trillions could be shifted virtually anywhere in a matter of nano-

★'While the First World races into the information age ... nearly 4 billion of the world's 5 billion people still lack the most basic access to simple telecommunications,' according to a report for the International Telecommunications Union (ITU) by consultants McKinsey & Co. cited in Frances Williams, 'Third World set to benefit from telecoms plan,' *Financial Times*, 3 February 1995.

seconds. As in a zoo where the animals are set free, the lions quickly devoured the lambs. London, New York, and Tokyo were consolidated as the world's preeminent 'money centers,' siphoning business from most small and medium-sized rivals. Communications played an important role. One large German bank now reckons that 'the deregulation of the British telecommunications market during the middle of the 1980s gave the financial center of London a head start over Frankfurt.'[2] At the same time, the globalization of certain industrial markets, and the resulting shakeout and consolidation of business into a handful of world players in each industry, also brought substantially higher telecom costs.

Playing their national card, industrial champions argued that to have any hope of surviving the cull and emerging as world players (like those free-roaming lions of finance) they would need an entirely new communications regime. And what better way to drive down costs, and at the same time encourage innovation, than to transform the public utilities into private enterprises and force them to compete? Governments obliged, starting with the 1984 privatization of British Telecom and the near-simultaneous breakup of the old Bell System. Within ten years this trickle had turned into a flood: virtually every European public telecom operator was headed for a stock-market debut. The goal was to transform 'networks ... designed to satisfy the nation states of the nineteenth century'³ into cosmopolitan competitors operating on a global scale. Freed to attack the market on bottom-line commercial terms, each of these freshly minted competitors has since been offering 'bulk' discounts, bespoke services, and sophisticated data management to entice large multinational clients into its fold. One result is that a company like Unilever, the Anglo-Dutch food and detergents giant, can now shop around for the optimal deal: 'virtual private networks,' an inexpensive way of coordinating entire organizations over vast distances, are available from all of the erstwhile telecom monopolists.

Of course, the benefits for individual homes and small businesses have been somewhat more mixed. Being geographically fixed, most still depend on the same source for basic telephone 'dial-tone.' Even a decade after privatization and liberalization, for instance, British Telecom still controls over 85 percent of the domestic UK 'retail' market. The pattern is similar elsewhere; as the result more of political decisions than of any 'unstoppable technical tide,' access to the communications network is being broken into multiple tiers, both within and among different countries.* A slow and not always dependable *public* Internet, for example, coexists with many fast, limited-access *corporate* nets. Instead of universal service, there are different services and different costs, determined by the market, not public policy.†

*An example of the new breed of player is MFS WorldCom, a new company whose August 1996 merger was made possible by the Telecommunications Act signed into law in the US earlier that year. The group is creating fiber-optic networks that weave together some forty-five money centers around the world. It will soon carry a substantial proportion of the world's financial traffic. This company already calls itself 'the first full provider of business services' dedicated to carrying such calls 'from the point of origination to the point of destination, internationally over a single company's facilities.' Thus control of a network and customers remains as

Words can change complexion in the most intriguing ways over time. Take, for instance, the word 'corporation.' Originally, it comes from the Latin *corpus*, meaning body, and *corporare*, to embody. However, by the late Roman Empire it had acquired almost spiritual overtones, suggesting a group of people united in a single belief. One Renaissance scribe wrote of 'the whole corpor-

decisive (from a business power perspective) as ever before: what has actually changed is the public-service responsibilities associated with that authority.

†In saturated (or low-growth) markets, there is often an incentive to try to squeeze out more revenue by offering *segmented*, value-added products. At the same time, in advanced telecommunications, there is an incentive for new competitors to offer services at lower rates, since they can piggyback on the existing cooperatively built network without having shared in the associated costs.

ation of those that are registered in the booke of life.' Today, of course, the word denotes something entirely different. It has been appropriated to mean a privately owned enterprise dedicated to the enrichment of the shareholders who control it. 'What we should term a Company of Proprietors,' Charles Dickens observed on a trip to the United States in the 1800s, 'they call in America a Corporation.'

This linguistic sleight of hand isn't always so imperceptible or slow: a word can change meaning almost overnight. For instance, when Wall Street stock-market analysts of the early 1990s talked about the *communications* industry, what they basically had in mind were the telephone companies. Cable and network television, plus newspapers and publishing, were off in another vaguely related sphere labeled 'media.' Computers were in a third world all their own. Now, cybernetic technology is reconfiguring this view with astonishing speed: it is driving all of these industries together, and, at the same time, it is raising the new industry's profile in relation to the economy at large. Communications will never be the same.

The technical reason is really quite simple. Originally, the telephone network was conceived as a voice-transportation *machine*; now, it is evolving into a high-speed *computer*.* This has major commercial implications. An American policy-maker hints at this with a sweep of his arm. We're seated in an office

*Electromagnetism, as elaborated by Alexander Graham Bell in the late nineteenth century, was what originally made it ↰

79

possible to send a waveform equivalent of your voice over long distance. As late as the 1980s, when you placed a call, electromagnetic signals traveled up and down through a hierarchy of local, trunk, and transit points, opening and closing switches in a vast meshing of mechanical instrumentation. Now the signaling equipment, the transmission pipes, the terminals, plus the software standards that control how they interact, have all been digitized. Wider streams of information (in computer code, not analog electrical signals) can pour down more capacious pipes. The plain old telephone is undergoing a metamorphosis, too, becoming an increasingly sophisticated microchip-driven 'terminal device.' This is transforming the cost structure of the entire business.

suite littered with electronic gadgets: a fax, a paging device, the phone, a laptop computer, and finally the ubiquitous telescreen. 'They're all digital,' he remarks. 'Soon you'll be able to plug into a cable network, a phone line, a satellite dish: whatever you please. And you can get anything: text, graphic images, movies − you name it. We can't fence one technology off from the next. We're not looking at a world of phone networks anymore: we're looking at a world of multiple interconnected networks.'

In that case, I ask him, what *happens* to all those phone companies, software houses, and TV stations?

'Well, essentially,' he says, 'they're all in the same game.'[4]

Under the burning sun of ancient Greece, the philosopher Heraclitus mused about the rushing river of life: it changes so fast, 'You can never step in the same river twice, for the waters are ever flowing past you.' Of course, nowadays the river seems more like a torrent. The great giants of communications hint that they're being swept to sea along with the rest, their organizational structures and business cultures crumbling away. Whole new business relativities are taking shape − new constellations of power. AT&T's chairman, Bob Allen, paints a dramatic, portentous picture. Screwing up his eyes and peering off into the middle distance like some grizzled captain confronting a storm, he sees only a 'boiling sea of change' and further struggles to 'reconfigure the enterprise.'[5]

*The leading global blocs (in order of size as measured by international call volumes) include AT&T/Unisource (itself a link between the public telecom incumbents of the Netherlands, Sweden and Switzerland), Sprint/Eunetcom (a link between Sprint, Deutsche Bundespost Telekom, and France Telecom), and finally Concert (British Telecom/MCI).

The challenges are very real. However, the incumbent telecom giants also enjoy a unique and even enviable advantage.* Until recently they were masters of a gigantic publicly regulated machine. Their market was mature, or 'saturated,' and therefore unlikely to grow. At the dawn of an information age, in contrast, they find themselves straddling what has been transformed into a vast computer – one that opens up a glittering frontier of potential new business opportunities. What's more (and despite all the talk about competing networks, 'superhighways,' and the Internet) their now-computerized net is still by far the dominant conduit for information exchange of all kinds, and will remain so for many years to come. And so – to the negotiating tables. Telecom companies are consummating new strategic relationships with the computer enterprises, the software giants, and the media conglomerates of this world in what can only be described as a dramatic combination of uneasy bet-hedging and clear-eyed intent.

'What you are seeing is the blurring of traditional barriers between various industries,' explains Pieter van Hoogstraten, strategy planner at Royal PTT Netherlands. 'You're seeing the growth of new clusters of business relations, where companies from different industries are interweaving and creating new forms of conglomerates in which communications plays the central role.'[6] So, notwithstanding that 'boiling sea of change,' many, including *The Economist*, foresee a 'revolution in communications that will change the globe.'[7] British Telecom's supreme leader, Iain Vallance, announces a new industry, poised on the cusp of 'the greatest growth curve the world has ever seen.'[8]

Three communications 'environments' are pictured over the next few pages. They are intended to show precisely how the physical landscapes of commerce are being segmented into exclusive domains. Success in the cybernetic economy will spring not from vigorous 'free-market competition' but rather from mastery of communication using systems and code. Mastery comes when three essential components – hardware networks, software environments, and customers – are successfully combined. The result will be new centers of decisive economic power that will operate in a networked, not a national, realm.

Big Business Systems

Imagine yourself in the warehouse canyons of Manhattan's lower West Side, near Houston Street. Before you, at 570 Washington, stands a featureless gray brick façade. This cavernous structure was once a rail terminus cum wholesale meat market. Today, fortress-like and austere, it shields the 'Network Operations Center' of Merrill Lynch, one of the greatest gods in the lofty pantheon of world finance.* Within, you pass under the glassy eye of the surveil-

*A financial services conglomerate with 500-plus offices in thirty-eight countries worldwide, Merrill Lynch had nearly $900 billion in assets under management in 1996. Its net revenue for the year was $13.1 billion.

lance camera, run a gauntlet of sober security checkpoints, then step into an immense industrial freight elevator and ascend three floors. You're ushered into the sanctum. It is a vast room, the nexus of an extraordinary global web that links the money and stock markets in every time zone with an army of traders and their many thousand investors worldwide.

Although Merrill, like other big multinationals, negotiates favorable rates on data-transmission capacity from local and long-distance telecom carriers, the trading system is an entity unto itself. Like a great moated castle, its data connections with the outside world reflect strategic alliances and tactical imperatives; they are strictly funneled along special channels, then checked past an elaborate series of gateways and secure software firewalls. Nested invisibly at its core lies an electronic operating 'environment.' This workspace, shaped by software (or code), is what drives the enterprise and forms the cutting edge of Merrill's thrust against global competitors. If this software can consistently deliver time-sensitive trading information to the right screen at the right moment, help execute a trade before the markets catch on, buying low and selling high, one day after the next, then the company will widen its already commanding lead.

It is an astounding operation. On one sprawling level, ranks of powerful mainframe computers are marshaled like so many tanks, row after serried row. The trading software has been written by former Pentagon propeller-heads using classified knowledge and advanced battle-simulation routines. Many kilometers of optical fiber and cabling converge on a 'bridge' whose glowing lights and sophisticated electronic modeling systems reveal Hollywood's Starship Enterprise as a tomfooling sham. This is real electronic war.* The consoles, crouched in multiple semicircular arrays, cybernetically catapult their operators worldwide. There is continuous and total surveillance. On this screen, to your left, you see a complex pictorial representation of the aggregate global traffic; on the next screen, to your right, the operator is executing a real-time diagnostics routine on a circuit card plugged deep inside a desktop terminal in

*See George Graham, 'War techniques may aid traders in City of London,' *Financial Times*, 1 July 1996, for a description of the great similarities between war zones and capital markets, and of the cooperative Financial Laboratory Club, led by the UK Defence Ministry's Defence Evaluation and Research Agency (patterned on its US-counterpart, Darpa), and involving leading UK investment houses and the London Stock Exchange.

Tokyo. A technician's fingers are dancing across the keyboard, rerouting a traffic overload. As you stand by, incomprehensibly dense pulses of electronic signal, magenta veins symbolizing billions of dollars of wealth, surge at her slightest command.

For Merrill and most sizable multinationals, efficient 'bitstream management' has become the essential prerequisite to business survival and growth. However, the costs soar with each necessary layer of added complexity: you need immense batteries of computing power and squadrons of rocket scientists and propeller-heads whose esoteric skills command (high) six-figure salaries. As the levels of complexity leapfrog, one past the next, and the costs spiral ever higher, all but the most mammoth players turn to outside help, thus surrendering elements of their informational sovereignty.[9] In a last phase, even the big players are swept into the net. The old power relativities have exploded, the fragments have begun once again to coalesce, and an electronically networked economy gradually starts to revolve around new hubs – alliances dedicated to synergizing their different centers, or 'nodes', of strategic expertise.

There was a time in this world of high finance when a fledgling newcomer, deprived of special discounts, advantageous relations, and the very newest of tools, could nevertheless bridge the gap between himself and his larger rivals by using native intelligence and hard work. No more. Even for the mid-sized operation the realities of survival are now touch and go. For, as Merrill's chief information officer admits, 'the days of the upstart are gone. It costs too much to stay at the top of this game.'*

*Author interview with Howard Sorgen, senior vice-president, Merrill Lynch, World Financial Center, New York City, 29 July 1994. In the US, financial services companies spend some $20 billion annually on upgrading their dealing-room and information-distribution technology. In the UK, the figure is about $3.24 billion. The fastest-growing sectors are real-time data-delivery and risk-management systems. Top vendors are Reuters, Dow Jones Telerate, Bloomberg and Knight-Ridder Financial. See Alan Cane, 'Computers in Finance Survey,' *Financial Times*, 15 November 1994, and Paul Taylor, 'Dealing room technology "at £2 billion",' *Financial Times*, 6 March 1994.

'Trading Places'

I've just arrived in Singapore on assignment. Surrounded by armored police vans and sultry tropical heat, waiting for a cab from the airport, I'm nervously fingering my press credentials. Buried in my bag is a copy of the *South China Morning Post*, which I picked up on the flight in from Hong Kong. It contains a disturbing report. Two senior Singapore government economists and the local newspaper editor have just been arrested in a fresh government crackdown on the press. They face trial for (believe it or not) disclosing their country's economic growth rate.

Singapore has taken a well-honed machete to all forms of unruly foliage, hacking it into the strictest submission. While the state's paternalist founder and self-styled 'senior minister' Lee Kuan Yew extols democracy, 'Singapore style,' the ruthlessly manicured gardens betray an obsessive program of control.[10] ('Singapore,' a T-shirt reads, 'is a really *fine* state.') Spitting on the street and dropping chewed gum are punishable crimes. News media are firmly muzzled. And yet this island-state's economic growth rate is the envy of Southeast Asia. Singapore's per-capita GNP exceeds that of Great Britain, its erstwhile colonizer, although for some reason publication of the good news is adjudged unacceptable. This is an ironic and remarkable combination – one that sends an ominous signal of things to come: here, we have cybernetic technology happily coexisting with the centralized political grip.

My visit coincides with Singapore's latest improvement drive: namely, to emerge as 'information hub' of the East, much as the cities of London, New York, and Tokyo now function as focuses of world finance.* TradeNet is a central weapon in this strategy – another code-based 'trading environment,' it was designed to link producers, shipping agents, freight forwarders, and regulatory agencies in a comprehensive 'vital community.' Say you're a local microchip manufacturer and you want to export electronic assemblies: your paperwork can be handled in

*A similar ambition is being pursued by Malaysia, which is building a $2 billion 'multi media super-corridor' in hopes of luring top IT talent and top infotech firms and thus vaulting from its medium-technology base to become an Asian leader in this field. It is offering tax holidays and other fiscal incentives to help realize this dream.

minutes on this system, while competitors in less wired cities are forced to wait days. Associated with one of the busiest ports in the world, the system enables shippers to track the location of their cargoes electronically, and helps them keep their customers abreast of any and all important changes. A vessel's arrival, berthing location, and departure are all managed cybernetically. Meanwhile, a strategic alliance has been established with Rotterdam, a freight-clearing center for the European hinterland, thus forming a competitive East–West 'axis.' For users of both ports, it all adds up to time and money saved: the system has fulfilled a deliberate design to advantage the axis in relation to rivals such as Hong Kong and Antwerp. As TradeNet's president observes, 'the name of the game is to stay one step ahead . . . if you are not on it [linked to the network], you are out of business.'[11]

A sobering thought. Mulling it over in the refreshingly jumbled order of home, my eye settles on a colorful bowl of fruits, glowing in a warm shaft of sun. My reverie is short-lived: I'm reminded that they, too, have been touched by these tools. As with flowers, their growth, sale, and delivery to market were managed by yet another trading network. The physical auctions, which in the case of flowers once brought all of the players together (plus busloads of tourists) to a massive trading-hall near Amsterdam's Schiphol Airport, are now being displaced by privately run trading places. Ditto with securities trading. Logistics and distribution. More and more, the broad trend is for physical, public markets that real people can actually *see* to be replaced by virtual, private exchanges where the action is invisible to all but a few with the appropriate screens.*

The entire corporate world is interconnecting electronically, with the computers that manage one company linking directly with computers in the next. For example a hospital will be *automatically* restocked with bandages, dressings, and ointments by its provider of medical supplies. A TV factory in Canton will be wired into its subcontractors in Bangkok; a car company into its distributors and dedicated retailers. John Malone, the boss of that stripped-down behemoth of cable television called TCI, conjures an image of corporate 'octopuses that will end up with hands in each others' pockets.'[12] It will be the electronic connections instead of the more familiar (and visible) cross-shareholding ties that determine the true nature of corporate relationships.† Meanwhile, the consumer will be captured in electronic 'malls' and other carefully designed 'environments.' The world is becoming a mesh of corporations and consumers interconnected less by physical than by electronic means. And it seems that companies failing to link with these wired networks are unlikely to survive, much less grow, into the twenty-first century. The reason, as Lotus's CEO John Landry explains, is that 'once those electronic links are established [between suppliers, producers, and their customers] it is going to be hell to get them out ... and try to get your own company back in.'[13]

*If this trend toward an increasing lack of transparency is allowed to continue without restraint, virtual markets will be far less troubled by the 'tiresome' burdens of regulatory oversight. Indeed, it is already hard to know *what* is moving *where*. When the London International Stock Exchange plugged in its on-line electronic trading system, in the 1980s, the floor was deserted in a matter of weeks. Visit the trading-floor of any European bourse today and you'd think it was dead: the deals are done in anonymous code. Similarly, the relationships between enterprises are embedded invisibly in electronic roads and gateways.

In an important study on global data flows, the Geneva-based International Telecommunications Union reports that official statistics shed no light on levels of encrypted data carried over private networks: 'It is difficult to know what portion of the world's data they carry.' Nevertheless, using the fragmentary figures volunteered by banking and transport industries, it hazards a guess that this river of private traffic has swelled at least threefold in less than a decade. The SWIFT interbank messaging system reports traffic growth from 157 million messages in 1985 to 457 million in 1993, and the SITA airline traffic network, which moved 2350 billion data bits in 1988, measured 6620 billion bits in 1992. SWIFT now faces a proliferation of competitors – including Ibos (the Interbank On-Line System) and Intuit – in a pattern being replicated in all sectors of commerce. (*Direction of Traffic 1994* (Geneva: ITU/Telegeography, Inc., 1994).)

'Consumer Communities'

Stretches of the California coastline are breathtaking. A century ago, this western shore was a glittering goal for pioneers and adventurers: America's last geographical frontier. Even today it conjures a sense of freedom from physical and cultural

†This is why these connections should be subject to reporting requirements, just as one company is obliged to reveal when it owns a part of another. Lacking this transparency, it becomes much harder to judge the relativities of market power and to ensure a healthy antitrust and competitive regime. (Antitrust is an essential tool whereby government seeks to prevent unlawful industrial combinations, monopolistic restraints, and unfair business practices, and to ensure that the marketplace is operating competitively and in the interests of the public at large.)

limitations, a place where you can reinvent a new identity. What better point of departure for the cyberneticists' *electronic* frontier?

As a matter of fact, I'm scheduled for an interview with one of these techno-visionaries at 3 p.m., so that all through my drive I'm torn between watching the clock and trying to soak in some of the still-untamed natural splendor that surrounds me. Suddenly, with a nightmarish shudder, the rental car winds down and stops. The ticking clock says quarter past two. Stepping out of my air-conditioned enclosure and onto the desolate, scrub-lined stretch of sweltering asphalt, I'm waving through plumes of steam billowing out from under the hood. What next? There's no sign of human habitation, and the last car passed ten miles back.

If this had been ten years ago instead of the wired California of today, the answer would have been simple: wait, hope, and pray. The interview would have been a washout, for sure. Now, miracle of miracles, I simply reach for a mobile phone. These handy communication terminals are plugged into a jack when you rent the car. They come equipped with a special red button which, when pressed in a situation like this, will instantly give you a connection to the Triple-A road service. At the same time, it will signal your precise whereabouts. The welcome outcome is that, thirty minutes and a bumpy tow-truck ride later, I'm at the corporate security desk, a bit disheveled but still ready to start my interview on time.

Maybe we should focus on this snazzy little red button for just a while: consider how it works; the lineup of business interests that brought it to market. The latest in an impressive string of modern gadgets, it's certainly hard to fault. It has saved me precious time and from missing an important exchange of ideas. And note: since the red button's been built into an otherwise standard mobile phone, I'm not *obliged* to use it — a person is welcome to consult the operator and hunt down the phone number of some alternative (and less wired) tow-service, if they please. But, then, this call is conveniently toll-free. The mere press of a button launches an almost fully automated procedure: more time and effort saved, which lets me focus on productive pursuits like planning my interview. And then, last and not least, I get the assurance of AAA's brand-name.

What it all adds up to is another driver opening up his wallet and joining a new (consumer) community. For the companies concerned — in this case the phone company (which has its network and the purpose-designed microchips), the rental-car agency (which 'delivers' its customer base), and the road club (with its web of service centers) — that little red button starts to look sweeter all the time. They've defined and captured a narrow but lucrative 'environment,' and will use it to widen their competitive lead.*

*The fundamental elements for corporate success in the cybernetic context are ownership of or favorable access to a communications network, control over a sufficient number of customers (or, more accurately, a combination of customers and their 'terminals'), and, finally, the appropriate software to establish connections between people and business groups.

This red button conveys the essential, and much wider, principle. If you know of a daily flight connecting St Maarten and the Caribbean island of Tortola, then you can track down a local travel agent and book a seat. But if you *don't* happen to know, and/or the airline's identification codes haven't been programmed to appear on (say) Air France or British Airways' reservation screens, the result is a phantom flight. If you're not in the software, you don't exist. Software is the decisive variable: it conditions the allowable interaction between individuals, trading environments, and big business systems. 'In an increasingly software-driven economy, design becomes *more* important rather than less,' maintains Derrick de Kerckhove, director of the McLuhan program at the University of Toronto. 'It practically *becomes* the content of the product.'[14] It's no surprise, then, that so much attention is paid to a software giant like Microsoft.

Most readers are generally aware of the background to Microsoft's breathtaking emergence in the market built on PCs. In the one year from 1993 to 1994, this company (founded by a man who once dreamed of collecting five cents from every car that passed a Seattle road crossing) rocketed from twenty-seventh place to become America's thirteenth biggest enterprise.[15] Visiting Microsoft's headquarters in Redmond, Washington, is very different from entering the consecrated halls of the old AT&T. Classical statuary is nowhere in evidence here. The first impression is one of invisible frenzy: work goes on round the clock. Cranes busily add to the haphazard collection of campus-like offices, linked by subterranean passageways. Microsoft's explosive sprawl is based on its *de facto* world monopoly in operating systems for personal computers.★ When the old, centralized power of mainframes was being splintered among countless distributed PCs, *software* was needed to make all those desktop units work (just as networks now rely on software to provide the glue of connectivity that keeps them up and running efficiently, one linked to the next).

Microsoft's proprietary operating system is of course MS-DOS, together with the accretions (like Windows) that have since been added on top. Intriguingly, its success had little to do with quality or cost: in fact, it never really had to compete. MS-DOS just happened to be available back in 1980 at a decisive moment when IBM

*Note that an operating system is the basic interface between a computer and its users: it determines how you communicate with the machine and precisely defines what it can (and cannot) achieve. Oddly enough, although we inhabit a computational world, the basic concept remains slightly puzzling to a great many folk. My parents and a surprising number of clever friends admit, sheepishly, that they're a bit hazy on the actual details. Yet they 'interface' with computers every day. Strictly speaking, the operating system is a collection of software programs that directly control the guts of your machine (or hardware). It is on this root system that all other higher-level programs – like word processing or communications or graphics – depend. Operating systems and application programs tend to be designed as a matched set – to run the latter and get things productively done you depend on the former.

was desperate finally to launch its belated drive into PCs. The deal itself was truly inspired: Bill Gates somehow sold IBM a license to use this cheap-and-cheerful system on the condition that a copy be packaged with *every* PC it sold.* The rest, so to say, is history. The PC, in effect, has since become a mere *wrapper*, or vessel, for the software it contains, just like most magazines are now vehicles for an ever-renewable stream of advertising. Riding on the back of IBM's market power, brand-name, and installed customer base, Windows has since crowded out all serious rivals to become the industry standard.† By the mid 1990s, if you want to work or play games on a PC, the chances are about five in six that you'll have a Microsoft application program riding on top of Windows, which in turn is built on the invisible MS-DOS base. You get the picture: a perfect, pyramidal hierarchy – the kind that information technology was supposed to destroy.

*Strictly speaking, the system was not at that time Gates's to sell: he returned from the IBM meeting and acquired the *ur*-DOS from an unknowing rival.

†In Windows, Microsoft 'borrowed' the icon-based interface and mouse combination that Apple had brought to market in the Macintosh. Except in its NT version, Windows is superimposed over the MS-DOS base – and needs it in order to work.

In effect, Microsoft chairman Bill Gates has successfully established personal hegemony over one of the emerging cybernetic domains. Like a feudal lord who straddles a strategic river crossing point, or a nineteenth-century railway baron with the most direct and well-connected rail lines, he's now in a position to *condition* the traffic to his advantage. For instance, he can pre-announce nonexistent products, sometimes thus discouraging competitors from striking out on their own. Then, by some mysterious form of osmosis, Microsoft allies (who write application programs that run on Windows) intuit the coming of technical changes with plenty of time to make necessary adjustments, and thus keep their own products working optimally fast. Rivals, meanwhile, are subtly wrongfooted and kept in the dark. And lastly, somehow or another, there's always this funny coincidence: Windows versions of hot software always materialize on the store shelves *before* those that run on Apple's Mac. This, of course, further dampens enthusiasm for the lowly but lovable Mac, whose

PC operating system was, for a time, the only serious rival on the consumer market to that produced by Mr Gates.[16]

Microsoft has built up an enormous critical mass of captive customers: almost anybody running a PC is using its code. Its software is comparable to the hardware infrastructure of telephony (or to certain industrial products like mobile phones and video recorders) in one crucial respect: the greater the number of people who use it, the more valuable it becomes. While Microsoft does not leverage its monopoly position to increase the price of existing products – that would smack too much of consumer gouging – each new iteration seems to increase in price. PC producers paid $33 for the old Windows 3.1 system but $43 for Windows 95. The new version of Windows, NT 4.0, will cost about $65 – or double the price of version 3.1. What's more, Microsoft's customer base enormously empowers the company when it comes to the negotiating table to talk strategic turkey with the big boys in media and telecommunications. It is in a prime position to deliver its 'users' to those wishing to sell them things.[17] Meanwhile, the next step is to link all those captive computers into a Microsoft network: a 'gateway' into cyberspace, running on the back of the feature-rich Windows 95. So, not only are you writing your letter according to Microsoft's Word, but also e-mailing that letter, organizing your personal finance, and even getting your news from a Microsoft publication and shopping in the Microsoft-sponsored mall on the Web.

This is the idea, anyway; in fact the business is evolving so fast that even the infallible Microsoft has had setbacks and been caught off guard. Nevertheless, the principle remains: code has become the adhesive that binds consumer communities and defines crucial electronic points of interconnection. As Chairman Gates observes, 'This new electronic world of the information highway will generate a higher volume of transactions than anything else to date; we're proposing that Windows be at the center, servicing all those transactions.'[18] And – needless to say – collecting that small toll.*

*The tactic is not unique to Microsoft. Compaq is offering to link any machine that contains a computer chip. 'We'll have an appliance that connects to a network for 6 cents a day. Compaq gets a piece of that 6 cents,' says a company executive quoted in *Fortune*, 1 April 1996.

Why does any of this matter? After all, these enterprises supposedly constitute the engine of electronic-age growth. Their dominance will rise and fall much faster than the physically bound empires of the industrial age. A company like Microsoft, triumphant today, might be displaced by an unknown coalition of gatekeepers in several years' time. (Recently, for example, a powerful allicance of computing, telecom, and software firms launched a scheme to 'migrate' much of the power, which is now controlled by the individual PC user into a network of centralized servers, which the user would access using a new, inexpensive, and low-capability 'Internet appliance'. These companies include Oracle, Sun Micro-systems, Netscape, and IBM as well as a number of leading US telecom firms.) These are early days in the Wired West: it, too, will settle into new standards and codes – and the new configurations of power associated with them.

But consider how much time we already spend on one network or another; how infrequently we visit the travel agent's actual office, or chat with a teller at the bank. Time is money – and these time-consuming human encounters are privileges for which we now have to pay at a premium rate. We spend far more time 'inter-acting' with software on a screen. Great areas of that vast continuum of human relations are moving away from physical inter-sections and into virtual 'sites': the focus of existence is shifting into a digitally mediated space. What this also means is that the sheer quantity of information is growing unmanageable, with data prolif-erating like kudzu strangler vines. Those with the skills to create and manage efficient electronic environments – those who can hack away the informational underbrush and deliver consistent informa-tional returns – are thus in terrible demand and rewarded accord-ingly. In short, those who assert management control over *points of interaction* will be most empowered in the cybernetic domains.

It takes some patience to understand the underlying dynamic, because the details almost always seem complex at first sight. (The reader is forewarned!) The following three paragraphs will take us into the lofty spheres of economic and monetary policymaking to try to clarify the relationship between coding and power. Now, one of the issues that is central to the success of the common

European Union currency – the 'euro' – is the structure of a centralized electronic payments system called Target. While European policymakers struggle with the issue of whether or not to lead their countries into a currency union, one of the key questions revolves around the terms and conditions of access to the Target system itself – decisions that will be fixed by technical rules embedded in code. It is the software architecture that will fix the ratio between the system's openness and its exclusivity and that will thus have profound competitive implications for the very balance of power in European finance.

Why?

Besides being a payments mechanism, Target will also serve as an instrument of monetary policy. The new European central bank – by pushing interest rates up or down or feeding greater or lesser quantities of the euro into the financial network – will use Target as an vehicle for the pursuit of a strong and stable common currency. Thus, officials from a monetarily conservative country like Germany might have reservations about allowing *unrestricted access* to the system by countries that have opted out from the currency union (and all the politically charged disciplines implied).

However, if Target eventually becomes the principal crossroad for large-scale payments exchange, then banks in countries that do *not* have full access to the system will be placed at a competitive disadvantage. This is because of yet another technically esoteric point, having to do with the dynamics of 'intraday liquidity' – which is the pattern of money flow between banks on any given day. Target will operate on the 'real-time gross settlement' principle. That means that payment transfers will be made instantaneously. (Today, in contrast, the many IOUs are settled on a net basis at the end of the day.) The implication? In a $1,200 billion-a-day foreign-exchange market, banks can quickly build up large and risky exposures to one another in a very short time, swinging from credit to debit in a flash. Now, to ensure that the banks can actually meet these obligations in 'real time,' and, simultaneously, to ensure the overall integrity of the system itself, it will be necessary for the banks to keep large euro balances at the central bank, or else be able to draw on overdraft facilities so as to meet their fluctuating needs. Therefore, if countries are positioned *outside* the monetary union's Target, and if their central banks are unable to issue overdraft liquidity denominated in euros, then banks from such countries (like Switzerland or the UK) will be forced to keep commensurately larger balances in cash. This, in turn, means that they will lose out on interest income they could otherwise have earned by putting that money to work. Result? They will be placed at a significant competitive disadvantage against rivals operating fully within the new regime.

This is an admittedly complex example of how human priorities, written into a system's software design, can be decisive in determining a competitive power balance. And if the issue of the euro is being decided by public policymakers, many other momentous decisions of social importance are being settled in the boardrooms of private enterprise and then quietly translated into code.

Such decisions will be decisive in a world where 'size [is] necessary for success ... but it is not sufficient,' as the *Financial Times* notes. For example, 'one of [News Corporation chairman Rupert Murdoch's] most lucrative skills to date has been to spot the "bottlenecks" or "gateways" in a market through which all participants must pass, paying the gatekeeper a fee as they do so.'[19] Similarly, on the Internet's World Wide Web, a fledgling electronic share-trading service called E★Trade, aimed at small investors, paid $15,000 in early 1996 for a slot on the *Wall Street Journal*'s Web site. It thus demonstrated that, even on the supposedly democratic and 'equalizing' landscape of the Internet, powers associated with a popular and convenient crossroads are proving as strong as if not stronger than before.[20] But the gates and bridges are digital, not physical, and thus they are harder to see unless you know what you're looking for. Even a *word* can serve as a crossroads of sorts. Say you're researching a topic like golf, telephones, or architecture. Now, because there is an unruly sea of available data on the Web, an efficient search requires *filters* of some kind. This is exactly what a Web-crawling 'search engine' does – it filters and thus manages data. Software producers can design these filters in such a way that advertisers can pay for the privilege of *sponsoring* an individual keyword. Whenever that word appears, the software can be programmed to behave in a specific way. Say the 'user' who has purchased that software fires it up and keys in the word 'golf'. The search result will be returned with a message from the sponsoring advertiser – who happens (surprise) to make golf clubs. Similarly, a communication company like AT&T or Sprint might buy *control* over the word 'telephone'. Thus a search for the word 'phone' will turn up ads for this firm. Such software can also display search results in such a way that advertisers' Web sites appear more prominently on the screen – even if they have less relevance to the user's concerns than those of other, non-sponsoring, sites.

The whole landscape of networked enterprise is being rendered subject to such forms of *signal management*, and it is this that will bring on new power relativities. Imagine yourself as the head of a Milanese architecture and design studio. Until a few short years ago your 'network' was a flexible web of human (or analog) contacts around the world. You did as much business in person as by phone. Your letters were dictated to secretaries, typed on gaily colored Olivettis, and then handed over to the tender mercies of the so-called Italian postal 'service'. Your designers had their proverbial drawing-boards, your mockups were modeled in clay, and your prototypes, whenever needed, were milled by hand and machine. You had ties with a wide range of suppliers – distinct industries boasting several competitors, anxious to supply you with letter bond, typewriters, drawing-boards, and modeling clay. If you were dissatisfied with one, it was easy to shift to another. Technology has now, by and large, scoured them from the scene. The configuration of your business – once independently tailored in analog space – is being swept into integrated environments where the hardware/software both defines and conditions the character of your relationships. Competitors who quickly make the shift into this new information space, and carve out their own proprietary territory, will reap benefits. They may well discover opportunities that were never open to them before, giving rise to all the talk of 'democracy and empowerment.' Once the canny Milanese architect is fully wired, for instance, she won't just *design* in cyberspace: she'll simultaneously *coordinate* her efforts with planners, contractors, and engineers. All of them, tethered together, will form a new virtual business community. (Meanwhile, the competitor who remains unwired won't even be able to *see* where this electronic alliance resides.) Thus the wired architect and her partners are not only empowered in relation to the laggards, who still insist on working with stand-alone phones and desktop PCs: if they are collectively strong enough, they will leverage their dominance over an important (information) crossroads as a competitive tool to discipline or eliminate rivals. The sheer power and speed of the tools will ensure that this advantage, once established, can be quickly consolidated into a commanding lead.

It is in this latter phase of the 'revolution' that the dangers will appear. Understanding the extent to which economic players can interconnect with each other, as well as the terms and conditions under which this takes place, will be crucial to recognizing the emerging landscape of economic relativities. This is why the invisible connections that are now being woven into the fabric of almost every commercial enterprise must be made more transparent. While such interconnections promise both to accelerate and to rationalize business procedures, and create altogether new markets and growth, they will also 'empower' those who make the physical-to-virtual shift. They may contain subtle biases that will limit opportunities for entrepreneurial initiative and the competitive spirit. They may ultimately lock participants into inflexible, unequal, and sometimes even unrecognized relations of business dependence. As commerce begins to revolve around large, communication-based hubs, the once unified information space could all too easily divide into new private spheres of influence, just as medieval Italy was carved up among warrior-kings. As an individual entrepreneur, your freedom of action within any such sphere will largely depend on how powerful you are. Those who enjoy the right economies of scale and scope – perhaps many of those networked corporations that even now control over a third of world GDP – will be powerfully placed to consolidate princely perquisites in the emerging electronic domains. Smaller companies, like mercenary knights, will also enjoy autonomy, especially in relation to non-wired competitors, but only on a more limited scale. After all, in the ecology of cybernetic competition, commercial survival is assured only when speed, size, and agility are optimally mixed.*

This phenomenon is not radically new – it is ageless. Only the context has changed. Throughout history, traffic has always been subject to some form of control: the central issue has always been what *form* the regulation was to take and who (or what) would perform the decisive intermediary role. Working as a correspondent in Rome, I once had a memorable interview with a man named Guido Rey which touched on precisely this point. A well-spoken public servant who was director of Italy's Central Institute of Statistics, Rey wrote a 1987 study that tried to capture the 'black' (or underground) economy in its estimates of Italian GDP. By the new measure published in his report, the Italian economy was much stronger than had previously been believed. The then premier, Bettino Craxi, was naturally overjoyed. A consummate politician, who later fled into exile from Italy in order to avoid multiple corruption charges, he seized on the report as evidence of his inspired leadership. His allies in the media were instructed to trumpet the news: Italy had surpassed the UK. '*Il sorpasso!*' the headlines screamed in a frenzy of self-congratulatory delight. In the midst of this operatic act, Guido Rey soberly received me in his spacious wood-paneled office, which lay some distance from the government palazzo. He wished

*A sharper distinction is likely to emerge between 'wholesale' and 'retail' electronic environments. In some cases, economic logic will dictate that environments be large: for instance, the Microsoft Network and Netscape Communications are even now building a mass of 'users' whom they can 'sell' to service providers in a wide variety of ways. In other cases there is a powerful incentive to maintain the scarcity value of specialized information, and therefore to keep the feedback loop as tight as possible. Thus, as we have seen, a handful of investment banks and specialized investment 'boutiques' can leverage the analytical expertise of their economists and traders with a unique and untouchably costly combination of software and hardware so that they can snatch crucial bits of financial data from the ether, place them in their correct context, and deal on the conclusions fractionally faster than anyone else. Those that lack sufficient resources to field these competitive tools and even costlier experts will be placed at a disadvantage and probably fall further behind market leaders.

to set the public record straight. In his judgment, the much-ballyhooed '*sorpasso*' was nothing more than a dismaying signal that something in the system was very much amiss.

'Back in the Middle Ages,' he began, 'if you wanted to cross the river Po and you couldn't afford a boat of your own, you had to pay the oarsman to pole you across. When the weather was fine and the boats were plentiful, it seemed as if all was well. When the weather was bad and the oarsmen were few, the situation would show its faults. The weakest passengers were always the most at risk. Standards of safety and service did not exist. Centuries later, we finally succeeded in creating a central government, and it raised money to build public bridges. *That*, you might think, was progress: everyone could cross the river on the same terms. Now, as my '*sorpasso*' report shows, the bridges have been allowed to fall into disrepair and those infernal bargemen seem to be back in our midst. Everyone is very excited. But ask yourself – really – is this such a good sign after all?'[21]

As Rey well understood, the medieval reality was one of feudal lords who ruled a fragmented landscape according to their own private whim. Allies were granted safe and inexpensive passage over their roads and rivers; enemies were decimated; all others were sub-jected to crippling tolls. It was only under democratic rule that industrial barons were obliged to conform to public discipline, to standards that were implemented in the hope of ensuring not only a diverse and competitive economic landscape, but also one in which opportunity was more equitably spread – like America's Ma Bell having to provide a phone service to all comers on similar terms. Now the entire public information space is being transformed into a territory for cybernetic colonization. Like the Americas, a brave new electronic world is being divvied up among the emerging kingdoms of twenty-first-century enterprise, with insufficient con-sideration for the welfare of natives or the inescapable issues of ethics and equity associated with its design.

Jorge Luis Borges, that Argentinean conjurer of visionary tales, was fascinated by the interplay between expectation and reality; by the intriguing way in which one's frame of reference can govern lived experience. This awareness lay behind his visceral aversion to mirrors.* The problem was not just that these demonic discs duplicated what (in his eyes) was already a vertiginous world: in one of his deceptively simple tales, a character is gripped by the horrified conviction that, if he ever once lets himself peer into a silvery disc, he will be confronted by a reality not just *mirrored* but subtly *transformed*. He was terrified at the idea of seeing his own face staring back at him with a separate intent.

The cybernetic world is one in which physical environments are replaced by an informationally based mirror. The reflective quality of this, the degree to which it reproduces the natural world, is powerfully conditioned by the humans who craft it as well as by the inherent, and very often invisible, attributes of the technology itself. When the world passes through the looking-glass, and when the replica displaces the naturally anchored reality to become the dominant terrain in which large numbers of people have situated their economic and cultural lives, then this mirror-image may suddenly take on independent, unexpected, and occasionally monstrous shapes.[22]

In the physical spaces of nature – in that awesome, chaotic, and endlessly flowing interplay of butterfly wings and distant typhoons – a release of one kind of energy triggers countless and ineffable permutations. In the cybernetically wired mirror world of a business network, in contrast, such infinite flexibility can never be attained. Unlike the human and natural spaces they're meant to replicate, these systems remain essentially *closed*, being both defined and bounded by artificial and *preselected* parameters. Freedom of action is a *programmed* variable. What you see,

*This metaphysical obsession with mirrors – and with the problem of illusion and reality – permeates all of Borges's work. Because of his near-blindness, he all but ceased writing stories after 1953 and concentrated instead on shorter works such as poetry and parables. One such parable, originally published in Buenos Aires in 1960, is 'The draped mirrors.' In it, he writes, 'As a child, I felt before large mirrors that same horror of a spectral duplication or multiplication of reality ... one of my persistent prayers to God and my guardian angel was that I not dream about mirrors. Sometimes I feared they might deviate from reality; other times I was afraid of seeing there my own face, disfigured by strange calamities.' See Jorge Luis Borges, *Dreamtigers*, trans. from *El Hacedor* ('The Maker') by Mildred Boyer and Harold Morland (Austin, University of Texas Press, 1964), p. 27.

what you can accomplish, the ease with which you can connect with others: all these factors are not determined by your individual will, nor chastened by the interplay of natural forces, but rather fixed by a *menu* of possible choices.* And, of course, the menu is written by Codemasters who have quicksilver priorities uppermost in mind.

*For example, if a critical mass of Web sites can be rendered technically incompatible with the older Web browsers used to access them, users will feel a subtle 'incentive' to migrate toward a newer, more feature-rich environment that has been 'adapted' to better meet the Webmaster's commercial needs and priorities.

Most readers already comprehend the basic idea of programming a device, if only from a consumerist (i.e. 'user's') point of view. We think of setting the video machine – or a coffeemaker. At most? Perhaps writing a little program to run on our PC. Masters of Code think in terms of entire companies, of virtual environments, and of whole cybernetic economies. They think of interlocking these systems into new and integrated spheres of influence. At present there are few limits to this development, save the ambition of the Codemasters themselves. So, like the Railroad Trust, the Sugar Trust, and the Cottonseed Oil Trusts that emerged from the frantic empire-building days of early industrialization, the stage is now set for an emergence of power in a new, communication-based mode.

[*Enter (stage right)* Cybernetic Trusts.]

Access to communication is a minimum prerequisite for participating in the information economy; the terms and conditions will decide how equably opportunity is spread. The reconfiguration of business along cybernetic lines therefore poses age-old questions in fresh terms. By what process will we hold the emerging Codemasters *accountable*? Can the public interest be safeguarded at a time when one technology forms the necessary locus of human interaction, yet the public information space is itself fragmented into countless private realms? Finally, a functional concern: as encrypted data traffic along the private networks of the very biggest players swells to floodlike proportions, will it even be possible for sovereign governments to monitor compliance with any eventual new laws?

Before the advent of cybernetic communications, business was woven into the fabric of society and subject to the discipline both of its leaders and of popular sentiment. Now it seems poised to transcend both on a global scale. The lords of cybernetic feudalism doubt the worth of policy intervention, and sometimes even portray 'big government' as the root of all evil. They claim that the quickening up/down cycles of the market make it impossible for anyone to stay on top for long; that technology *protects* the innovator and acts as a natural equalizer. In fact they seem set to shake off that 'heavy hand' of the law itself, and to elude the public responsibilities that prevailed in the analog domain.★ In an information landscape that is digital, not national (as one antitrust regulator privately confesses), 'a very real concern is that we'll end up with a few pivotal organizations that are so powerful in their spheres of influence that they can simply say, "OK, these are the rules."'[23]

British Telecom's Iain Vallance has declared that these 'electronic networks are like the nervous system binding modern society.' What's more, he notes, they interact in ways that are 'frightfully complex.'

'So, what of public policy regulation?' one inquires.

★In a high-bandwidth, data-saturated economy, strategic information resources with the highest relevance and scarcity value will be gathered toward increasingly inaccessible cores. Editors, or 'information managers,' will have a commensurately more powerful role. Among the critical unanswered questions is how to guarantee fair access to information, and to the conduits for its exchange, in an environment where the once unified public network has been fragmented into a series of interconnected but proprietary private nets. Other questions include how to measure and control excess market power by any of the emerging participants, and how to gauge the true nature of their alliances, which are no longer accurately reflected in their cross-shareholding ties.

An odd smile plays about his lips as he replies, 'Regulation? Perhaps that is beyond *anyone*'s capability.'[24]

5
Masters of Code

... the custom ... of congregating together for mutual protection is still kept up in most parts of Spain, in consequence of the maraudings of roving freebooters.

WASHINGTON IRVING[1]

In a preternatural fall glow, I find myself strolling through the City of London, a pulsing electronic node in the networked world of financial enchantment. Gazing up at the gilded cap of one of its office towers, an Egyptian image unexpectedly springs to mind. I recall that, somewhere around 2600 BC, the great pyramid of Cheops at Giza was also (proverbially) capped with gold, and its surface sheathed in a highly polished white limestone shell. When Herodotus visited it in 440 BC, more than 2,000 years later, he reported that the structure still had such an extraordinarily smooth finish that it was difficult to discern just where the stones were joined. Archaeologists now believe that once, beneath the pyramid's glassy surface, outlandish rites were orchestrated by a hermetic and all-powerful priesthood.

Maybe these pyramids are not so distant, in this respect, from our own pharaonic (if structurally flimsier) pillars of high finance. London, New York, Frankfurt, Tokyo: they're all the same. Like the pyramids at Giza, their skyscrapers convey an aura of imposing if mysterious purpose. From the sidewalk looking up, a blank face of pale mirrored glass. Somewhere inside are windowless bunkers swarming with snake-like coils of optic fiber and wire. Traders offer odd incantations to computer consoles – which magically project arcane electronic signals to wizardly counterparts a continent away. Fortunes are made: nothing is produced. The uninitiated mind is left to reel at this strange link between abstract information and monetary value.

Business – in the post-Cold War environment – has become a new form of (sublimated) global war. And information itself – together with the tools for processing that information – has become the predominant weapon of battle. This is in keeping with the history of war and weapons: from earliest days until today, the distance between the protagonists and the consequences of their acts has steadily increased. For instance, when a medieval knight hacked with a battle-ax, the result was a very visible gash in his adversary. Gradually, however, the days of the sword and lance and single combat started to give way to those of arrows, rifles, machine-guns, and guided rockets, until finally, today, warfare is conducted at a sanitized technical remove. The consequences of battle, prosecuted by electronic proxy, are invisible to those who control the triggers. Not long ago, a pair of UK traffic policemen almost found themselves on the receiving end of such electronic proxy power when they were out ensnaring motorists with their radar gun. Quite unexpectedly, the radar gun registered a seemingly impossible speed. One millisecond later, as a Harrier fighter jet screamed overhead, their digital readout went berserk, and numbers started twitching wildly across the screen. The officers later phoned the RAF to complain about the damage to their equipment. A squadron leader let slip that they were actually rather lucky to be alive. (It seems that the Harrier jet's target-finder had mistakenly locked onto the officers' 'enemy' radar and instantly triggered a retaliatory air-to-surface missile attack. Fortunately for the two constables, the Harrier was flying unarmed on that particular day.)

Digital weaponry has enabled us further to distance and objectify commercial as well as military targets. From a cybernetic cockpit, those who have obtained political clearance for takeoff can proceed to project their financial ambitions over the landscape at large. Indeed, their digital systems are surprising comparable to those that propel the Harrier jump-jet. They exist in the interstices of silicon and code. They have been programmed to operate in pursuit of numerical abstractions. Increasingly, they are so quick and efficient at pursuing those targets that no lone pilot is capable of manual override – for such an override would require a complete overhaul of the system's design and aims. Meanwhile, lesser mortals appear on the screen as a collection of blips to be automatically identified as friend or foe. When the digital economy targets the analog business networks and their stakeholders in tangible space, it is the ordinary souls who get blown away.

Ordinary mortals in the post–Cold War world now confront a profound new danger from within. It comes in the shape of a narrowly based, unrestrained, and deregulated financial order that is being powerfully projected by technical means. It is leapfrogging the essential and restraining networks of local accountability; it is undermining the essential trust on which economic vitality depends. Many still wonder how this came to pass. Its origins lie in a re-structuring of the financial system that was put in place at the end of the Second World War, when the US dollar had firmly dis-placed sterling as the preeminent currency of world trade. All other national currencies revolved around it, and its value was pegged on gold bullion. This pegging of the dollar's value to a physical com-modity seemed to keep exchange-rate fluctuations to a minimum: they played little if any role in the long-term investment strategies of enterprise. When bankers lent money to business, it was turned into productive plants and wage-paying jobs; it was used to trade in real goods and services. Business and finance were intertwined in a relationship of mutual interdependence – and their game was supervised by government referees.

In the 1970s the picture changed. America had created a problem for itself by over-issuing dollars to finance its Cold War defense buildup, its Vietnam adventure, and its overexpansive commercial thrust abroad. There were suddenly far more dollars floating around the world financial system than there was actual gold to redeem them. Also emerging was the technical threat that they might now move in potentially debilitating ways. So, in August 1971, US president Richard Nixon expediently abandoned the gold standard, scrapped the postwar Bretton Woods order, and ushered in the roiling chaos that one practitioner has since christened a 'floating *non*-system.'[2]

This decision, coupled with various deregulatory 'big bangs' and the gradual lifting of foreign-exchange controls a decade later, gave birth to virtual finance. As more and more money came to be floating around – beyond the reach of traditional regulatory restraints – the disciplines of banking and business inevitably began to diverge. High-speed cybernetic communication devices made it possible to shift ever greater sums of digital 'cash' at the blink of an eye. The market was awash with funds. Cash-rich pension assets were joined by so-called mutual funds, in which small investors began to invest their savings in the expectation of quick returns. Speculative expectations further inflated prices. More electronic funds chased a limited number of tangible, real-world assets. There was suddenly more money to be made in moving bits and bytes than in producing real things and generating jobs. The nominal wealth of the financiers spiraled ever higher. But performance pressures on real-world companies – competing for investment and trying to match these inflated returns – have forced them into increasingly harsh expedients. They have trimmed their labor pools, driven down salaries and benefits, and sought computational efficiencies to stay in the competitive race. Even the most well-meaning chief executive sees little choice: meet the quota or join the dole queue with the rest. As a recent US labor secretary, Robert Reich, remarks, 'Gone forever is the gentlemanly investment system that once allowed chief executives to balance the interests of shareholders against those of employees and communities. Now, chief executives who *don't* abandon their employees and communities when the bottom line requires it risk trouble – while those who *do* are able to pocket multi-million-dollar bonuses and stock options.'[3]

Although it is never reflected in any existing profit-and-loss statement, the true cost of this unrestrained and electronically distributed race is that the close relationships that once obtained in the world's financial centers are breaking down. According to one recent report, commissioned by the London Stock Exchange, 'If you can make a killing in one successful deal, you may not be so worried if in the process you weaken bonds of trust.'* The sums and incentives attached to this game are indeed immense. Turnover on the currency markets (never mind stocks, bonds, or derivatives) now approaches a mind-reeling $1,230 billion per day.[4] This is far more than most countries' total national output of goods and services every year. As electrified money cycles through cyberspace in faster permutations of ever-increasing complexity and abstraction, it is ultimately decoupled from economic activity. It follows reflexive impulses of its own. The financial markets have been transformed into arcades devoted almost entirely to speculative exchange. They are now largely run by computers that are programmed to achieve certain rates of return, measured as a percentage of the sums at risk, by whatever are the most expedient means. Writes Joel Kurtzman, a consultant and former editor of the *Harvard Business Review*:

> ... computer programs are not trading stocks, at least in the old sense, because they have no regard for the company that issues the equity. And they are not trading bonds per se because the programs couldn't care less if they are lending money to Washington, London, or Paris. They are not trading currencies, either, since the currencies the programs buy and sell are simply monies to be turned over in order to gain a certain rate of return. And they are not trading futures products. The futures markets are only convenient places to shop. The computers are simply making transactions. They are moving tokens across the megabyte economy ... they are trading ... ghostly images that symbolize buying power, images that store labor, wisdom, and wealth the way a computer stores numbers ... [they are] trading mathematically precise descriptions of financial products ... any one will do if its volatility,

*Deregulation and competition narrowed profit margins in traditional bank business, while computerized global financial flows and the growth of derivatives markets made it cheaper for banks to generate income from trading, which had until then been dominated by securities houses. Describing the City of London, one highly regarded business journalist remarks that 'No other financial center in history has enjoyed such success while being semi-detached from the domestic economy.' He adds, 'The growth of telecommunications, far from leading to a dispersal of financial services, appears to have encouraged centralization by allowing London to serve a wider hinterland not only in the European time zone, but across the world.' (John Plender, 'Still growing after all these years,' *Financial Times*, 13 March 1995.)

price, exchange rules, yield, and [risk] fit the computer's description. The computer hardly cares if the stock is IBM or Disney or MCI. The computer does not care whether the company makes nuclear bombs, reactors, or medicine. It does not care whether it has plants in North Carolina or South Africa.[5]

Under the influence of networking and deregulation, an economic world built on patiently cultivating inherent worth has given way to one devoted to immediate numerical 'growth.'* This system is volatile, breathtakingly profitable (at least for its managers), and has been responsible for a mass redistribution of global wealth. The Mexican peso crisis was one of the more notable examples of the dysfunctional dynamics involved:

> For years, Mexico increased its foreign borrowing ... including selling high-risk, high-interest bonds to foreigners; selling public corporations to private foreign interests; and attracting foreign money with the speculative binge that sent [its] stock market sky-rocketing. As little as 10 percent of the some $70 billion in foreign 'investment' funds that flowed into Mexico ... actually went into the creation of capital goods to expand productive capacity and thereby create a capacity for repayment. [In consequence,] projected debt service payments alone came to exceed ... export

*The complexity which cybernetic technologies have brought to modern financial operations can be hard to grasp, even for initiates. Small amounts of actual capital can be leveraged to create large deals with many interlocking variables, each affecting the rest. These transactions are of such exponential intricacy that in many cases they are not fully understood by the companies that promote them. Only computers can design, manage, and profitably close these deals. However, there are serious questions about how accurately computers can measure elaborate correlations (and thus give a reliable picture of risk) at times of irrational, extreme, or particularly rapid market fluctuation. Even before the fantastic elaboration of today's derivative markets, an interdisciplinary group including Citibank's John Reed met under the auspices of the Santa Fe Institute at Rancho Encantado in New Mexico on 6–7 August 1986 to consider 'Evolutionary paths of the global economy.' The report concluded that 'The world financial system is complex, multi-centered, subject to severe disturbances or even breakdowns, and poorly understood in its more dynamic aspects.' (The proceedings of the interdisciplinary meeting were published as *The Economy as an Evolving Complex System*, ed. Philip W. Anderson, Kenneth J. Arrow, and David Pines (Menlo Park, Cal.: Addison-Wesley, 1988), which comprised vol. V of the Santa Fe Institute's Studies in the Sciences of Complexity. The text above appears in the appendix to that book.)

revenues. Mexico's 'economic miracle' [proved to be] a giant Ponzi scheme.*

*Charles Ponzi was an Italian-American swindler of this century who paid initial investors with money from later investors who hoped to receive the same fabulous returns. The quote is from David Korten, *When Corporations Rule the World* (London: Earthscan, 1995), Chapter 3, note 23.

When the bubble finally burst, in December 1994, Mexico was blessed with twenty-four billionaires where it had previously boasted only fourteen. US and European taxpayers had meanwhile funded a $50 billion scheme to ensure that Wall Street and City of London banks and investment houses recovered the bulk of the money they had staked on the speculative game. In early 1996 the bailout was extended to cover a number of large Mexican corporations.[6] On the other side of the ledger, some 4 million Mexicans (or 10 percent of the population) were forced into short-time work involving less than fifteen hours a week. A further 800,000 lost their jobs altogether. The final blow came with a fourfold increase in interest rates – calculated to reverse the frightened outflow of foreign investment. This had the result of sending scores of ordinary families, who had also staked a claim on Mexico's future, into direct insolvency.†

Such dynamics cannot be dismissed as mere aberrations: they are the rule. In the decade during which electronic finance gathered

†See Leslie Crawford, 'Mexico's vigil of woe,' *Financial Times*, 2 June 1995; Damian Fraser, 'Bishop at the heart of the struggle,' *Financial Times*, 18 January 1995; Leslie Crawford, 'High on the danger list in Mexico: Ailing companies queue up for special government help to avoid bankruptcy,' *Financial Times*, 24 January 1996. Not coincidentally, this period of 'adjustment' was accompanied by a rebellion of native Indians in the southeastern Mexican province of Chiapas: an uprising triggered by a desire to assert their autonomy amid foreign takeover of local lands and an imposition of alien economic lifestyles. The rebellion, notable for the scanty and uneven coverage it has received by the press in the developed world, still simmers to this day.

pace – between 1982 and 1993 – the listed net worth of *Forbes*'s 400 richest people in America swelled by $92 billion to a total of $328 billion. For purposes of comparison, this figure exceeds the total gross national product of the entire Indian subcontinent with its populace of 1 billion souls. During the same period, the average income of most US citizens fell substantially, in both real and absolute terms.* The world as a whole now has 358 men and women with declared assets in excess of $1 billion. Combined, their wealth exceeds the total annual income of half the planet's population.† Meanwhile, the flow of bits and bytes goes on, rendering certain modes of life 'uneconomic' (as measured in strictly financial terms), while favoring the efficiencies that only networked enterprise can achieve. One result, claims John Perry Barlow, a prominent cyberspace libertarian who recently authored a portentous document entitled *A Declaration of the Independence of Cyberspace*, is that 'if you are making something you can touch, and doing well at it, then you are either an Asian or a machine.'‡

Thus the benefits of electronic finance, as with commerce or electronic battle, accrue to those who control the systems that manage it. In effect, a distant overlay culture is mobilizing the tools of global networking to remotely extract wealth from the various 'nodes' tethered to the net: local relativities and

*'Between 1973 and 1993 the real hourly wage of Americans without a high school diploma actually fell from $11.85 an hour to $8.64 an hour. In the early 1970s households in the top 5 percent of the income bracket earned ten times more than those in the bottom 5 percent; today they make almost fifteen times more.' This unsustainable polarity must be eased, suggests Ethan Kapstein. ('Workers and the World Economy,' *Foreign Affairs*, vol. 75, no. 3, May/June 1996.)

†*Human Development Report* (New York: United Nations Development Programme, 1996). Over the past thirty years, the report notes, the poorest 20 percent of the world's population saw their share of the global income drop from 2.3 percent to 1.4 percent, while the share of the richest 20 percent rose from 70 percent to 85 percent. This growing divide between rich and poor, if continued, will yield a world even more 'gargantuan in its excesses and grotesque in its human and economic inequalities.'

‡'Wearing a bandanna, blue jeans, boots, and the rugged aura of an outdoorsman instead of the pressed stodginess of a corporate soldier, Barlow both literally and philosophically represents individualism. His multidimensional character bears no resemblance to anyone likely to be found in the corner offices and boardrooms of Corporate America. Or does it?' asks David Bottoms in 'Cyber-cowboy ... or prophet?'(*Industry Week*, vol. 244, no. 22, June 1995). Barlow, who describes himself as America's only Republican hippie mystic, is a former lyricist with the rock band The Grateful Dead, a one-time cattle rancher in the hills of Wyoming, and co-founder of a lobbying group called the Electronic Frontier Foundation (EFF). In his latest incarnation, he acts as a business consultant (as befits one of the more articulate of the cyberspace libertarians) and a digital guru to the great unwashed public at large. The quote above was downloaded from the Barlow Archive at http://www.eff.org. In his 'Declaration ... ' Barlow characterizes governments as 'weary giants of flesh and steel,' and asserts that they have no legitimate authority over the informational realm.

lifestyles are being subjected to the uniform tyranny of their inflexible, placeless, and quantitative demands.*

*Of course, even today many localized markets with independent means of exchange persist within larger structures; the underground and drug economies are good examples.

†The term was coined by Tom Wolfe in his satirical novel *The Bonfire of the Vanities* (New York: Farrar, Straus & Giroux, 1990).

Walter Wriston is a former Master of the Universe† who occupied the chairman's suite at Citicorp for many years and was recognized as being one of the 'most competitive [of] bankers ... a driving profit seeker who saw the world as his battleground.' Wriston's ride on the roller-coaster of wired finance – which he once described as 'trying to deal with a succession of accidents'[7] – strengthened his conviction that 'technology has begun to bypass politics ... [and that] no matter what political leaders do or say, the screens will continue to light up, traders will trade, and currency values will continue to be set, not by sovereign government but by global plebiscite.'[8]‡

Lacking funds, advanced tools, and wizardly expertise, ordinary citizens and their priorities are rendered irrelevant. They are excluded from the arena in which decisions are made. The sheer size, immediacy, and power of the financial market – a 'plebiscite'

‡For all the liberal market and anti-government rhetoric of the banking elite, the stabilization of the Third World Debt Crisis and similar self-inflicted wounds depended crucially upon taxpayer-financed intervention by sovereign governments and international authorities. Moreover, three years after Wriston retired from Citibank, his successor, John Reed, was forced to set aside $3 billion to cover loan losses in Latin America, corresponding to the total of the bank's earnings during Wriston's last four years of service. Soon after that, the bank nearly collapsed

in which only money may vote – are imposing what are essentially *political* choices on civil society. John Gilmore, a cybernetic guru, is famous for coining a phrase that 'the Net interprets censorship as *damage* and routes [information] around it.'[9] And networked world finance has come to regard *public policy* as a form of censorship too.§

under the weight of bad loans in commercial real estate, but was able to avoid a stampede of customer withdrawals because of the existence of government deposit guarantees. Banking, being essentially a high-risk business, cannot survive without the safety net and the stabilizing restraints that only a government can provide.

§For example, the logic of European Monetary Union (EMU) is set very much against the background of the imperatives of competition in a networked global economy. In theory, the introduction of a common currency, called the 'euro', will mark a definitive transformation of Europe from a collection of closely associated states into the world's strongest integrated economy, whose monetary policy will be coordinated by a single central bank based in Frankfurt. The hope is that this euro will emerge as a formidable competitor to the dollar as the currency of international trade. However, the EMU agenda has been poorly explained to the public at large. It also requires a harmonization of varying national tax and social spending policies, each reflecting a different mix of cultural and historical relativities, and thus will also subtly and rapidly strip national electorates of a substantial degree of control over their own collective destinies. National authorities are not being replaced by sufficiently accountable supra-national representative structures that can exert countervailing democratic control. Thus, as presently designed, EMU is an agenda whereby the unified power of money will evade that of fragmented electorates.

This can neither be avoided nor controlled, Wriston maintains:

> Even though markets are now blips on a screen and not geographic locations, sovereigns still try to protect and control that part of the market that functions within its jurisdiction. Yet even this becomes increasingly difficult, for if one sovereign becomes unreasonable in the severity of regulatory demands, the market *node* in that country withers and is replaced by the node 'residing' in more hospitable climes ... in an information economy, government rent [tax] collectors have much less leverage because *the tenants can leave town*.[10]

Alain Levy-Lang, the outspoken chairman of France's Banque Paribas, underscores the point: this free-form 'global capital market [has become] ... *the* controlling factor in national economic policy.'[11] People's lives and job prospects, their hopes for the future (and that of their children), are now said to lie entirely in the sprawling electronic markets' embrace.

What can be done? Peter Drucker, one of the great exponents of modern management theory and a guru of industrialists worldwide, insists that the stark and inevitable choice in the current restructuring is either to kill or else to be killed.[12] Of course, informational weaponry eases the moral strain by transforming the human casualties into invisible video blips. The protagonists can speak of imperatives – and preach the competitive ethic from their lofty altars – and present hands that seem perfectly clean. They are the mandarins of ultimate pragmatism. In his great tome *The Post-Capitalist Society*, a bible for information-age executives, Drucker writes, 'Society, community, family are all conserving institutions.

They try to maintain stability and to prevent, or at least slow down, change. But the organization of the post-capitalist society ... is a destabilizer ... its function is to put knowledge to work ... it *must* be organized for constant change.'[13]

Thus, in a wired economy, living conservatively – at least in the classical sense of the term – is tantamount to living in sin. As Hannah Arendt has famously proclaimed, 'a booming prosperity ... feeds not on ... anything stable and given but on the process of production and consumption itself ... not destruction but *conservation* spells ruin ... constant gain in speed is the only constancy left wherever it has taken hold.'[14] Even the business chief executive, who labors mightily to expand his corporate sales base and streamline his efficiency, can suffer an instant and irrational reverse if the currency markets move the wrong way, or if he fails to deliver sufficient levels of short-term growth at whatever long-term cost. He feels obliged to set aside personal values and priorities to satisfy the market's voracious demands. Meanwhile, as Italy's Roberto Calasso writes, 'the Pauper ... who wants only to continue existing, who desires his land more than the development of his land, the "wretched ingrate" who spurns the offer of money: they are an outrage and an affront ... the signal of the fierce revolt of the land against its hasty, impatient surveyor.'[15] For Calasso's 'hasty surveyor,' the mandarin of a global overlay culture that claims to act on behalf of society at large, the source of wealth is a safely distanced abstraction. Only the others – those 'wretched ingrates' on the ground – fully recognize how cataclysmic this unrestrained movement of bits and bytes can be.[16]

In a very real sense, the wizards of electronic finance have captured the key to the social temple. They have assumed the freedom to strip the altars of all their finery, and a mandate to transfer the realized value to the credit of their own accounts. They strut from the ruined sanctuaries – never having slipped off their shoes – and race after the sun with their electronic funds.* When the financial markets close down in Tokyo, the action shifts to London and then on to New York City. Speculative trading – much of it off bank balance sheets and/or otherwise

*In 1992, by betting $10 billion against the pound sterling's valuation within the European exchange-rate mechanism, George Soros's Quantum Fund successfully scrapped the exchange-rate system itself and emerged with

a $1 billion profit. The general public pays a high price irrespective whether such gambles succeed or fail. For example, unwinding the speculative excesses of the underregulated 'free market' in the case

opaque – plus commissions on simple turnover, now accounts for the bulk of the finance sector's profit.

Movement – in this age of digital mobility – is the message.

The 'free-form' markets were created not by technology alone, but rather by a unique *interaction*

of the US savings and loan 'accident' required a staggering $145 billion capital infusion, of which $120 billion was provided by American taxpayers. A similar bailout took place in Japan in early 1996, involving some $20 billion in taxpayer receipts. (See Gerard Baker, 'Japan approves more public funds for housing loan bailout,' *Financial Times*, 29 January 1995.) Too often, in underregulated 'free markets,' short-term profit has been extracted by a few at a long-term cost borne by the many.

between technology and political decision. In their self-absorption, the beneficiaries forget that it was a presidential abandonment of the gold standard, followed by progressive bouts of deregulation in leading financial markets, that created the preconditions in which their arcade game could actually be played. If markets and financiers are now connected in a self-perpetuating chain of global risk, it is only because this has been sanctioned by policymakers and tolerated by electorates to date. In their amnesia about this deregulatory largess, the winners seem to be developing a certain arrogance, to say the least. In 1994, when US president Bill Clinton proposed to tax futures and options trading in hopes of dampening market volatility, Jack Sandner, an official of the Chicago Mercantile Exchange, issued a public statement informing the president that, if he were to do so, business would move overseas 'in a nanosecond.'[17]* The president of the United States was also reprimanded

by an employee of the Chicago Board of Trade – an individual by the name of Patrick Arbor – who said that merely floating the possibility of a transaction tax 'seriously jeopardized' the industry's 'fragile

*Under the current regulatory framework, a government can still influence the financial markets by its tax, spending, and interest-rate policies, but will be punished if it strays too far from norms set in the boardroom and among trading-desks.

competitive advantage worldwide.'[18] The president subsequently caved in and dropped the plan.

The potential benefits of such regulatory arbitrage – a process of playing one government off against the next – are hardly limited to finance. By fragmenting and accelerating the financial game, informational networking tools have supposedly created what Banque Paribas's Levy-Lang calls 'a fundamental contradiction between ... labor rules, environmental standards and tax structures on the one hand, and the requirements of industrial competitiveness on the other.'* In a powerfully networked economy without a correspondingly aggressive political counterweight, the tendency is for standards to be harmonized down to the lowest common denominator: toward short-term expedients that favor a privileged few and shift the associated long-term costs to the many.[19]

*This inherent and unresolved tension between the abstract demands of electronically mobile capital on the one hand and tangible political, social, and economic aspirations on the other already lies at the heart of some of the most divisive issues crowding onto the political agendas of our day. Although this tension is often presented as unprecedented and insoluble – an accomplished technical reality – it is more properly interpreted in the timeless terms of ethics, accountability, and political will.

For instance, in January 1995 the petrochemical giant Royal Dutch Shell decided that, in order to avoid paying UK National Insurance contributions for its British seafarers, it would simply transfer their contracts to Singapore and administer them from the Isle of Man. This saved Shell 40 percent of its previous payroll cost. Meanwhile, after the 1989 *Exxon Valdez* disaster, the 1993 sinking of the tanker *Braer* off the Shetland coast, and a large number of other destructive oil spills and assorted safety abuses, the United States proposed in 1994 to upgrade oil-tanker financial-liability standards. (Shell was again cited, because one of its gas tankers was found with four-inch cracks in its deck after a rough Atlantic crossing – an incident that one company spokesperson described as 'very minor stuff.'[20]) In 1993 no fewer than 121 ships were lost at sea, at a cost of 592 mariners' lives – an increase of 35 percent on the year before. Despite the fact that one in five of the world's tankers is operating below international standards, and some 70 percent of all tankers are already registered in countries that offer weak regulation, to save owners' costs, the oil majors responded to the US proposal with a blunt threat that either they would 'reflag' (that is, re-register) their vessels elsewhere – a move which would cut US jobs and also further erode the vestiges of US regulatory control – or, alternatively, they would simply refuse to sail in American waters. The safety-improvement demands were subsequently diluted, leaving an asymmetrical patchwork of regional rules which form a loophole through which owners continue to sail.[21]

Many individual governments are effectively abandoning independent social and economic policies in exchange for the future benefits that a fully wired world of digital laissez-faire will supposedly bring. Arguably, however, inaction will ensure that these benefits never materialize: the human and ecological costs of unrestricted chaos will prove too great.

Since precious metal coins were invented by the kings of Lydia in the seventh century BC, and through to the development of paper currency 2,400 years later, riches were accumulated through trade and investment. Control of the money supply depended upon either the discovery of new wealth in mines or the reminting (and debasing) of actual coins. The more the coins were debased, the further their buying power fell. The same holds true for governmentally backed paper, which requires strict economic discipline and regulatory controls: lacking these, the value is marked down. At the extreme, the result is hyperinflation and depression. But now, entirely abstract forces have come into play, creating or destroying wealth by leveraging the power of informational tools without taking sufficient account of real-life consequences. Already, for every single dollar spent or invested in the 'real' economy, an estimated $30–50 are floating in financial cyberspace.[22]

Soon, consumers will have access to *privately minted* digital dollars, or e-cash, made available by broad-based commercial alliances revolving around hubs of financial, communications, and industrial expertise. Consortium-backed electronic money will significantly enhance the power of private rather than publicly regulated networks of wealth.* Governments will find themselves competing directly with commercially issued currencies. More of the money supply will be effectively 'privatized.' The electorate's leverage over its own social and economic policy

*This important idea is further explored in Chapter 8. The consortia are hoping to build on their brand-name recognition to displace government-guaranteed currency with their own, privately issued, electronic cash. They ⟳

initially hope to capture a sizable part of those consumer transactions in which less than $10 changes hands – over $1.8 trillion worldwide. This is tantamount to 'privatizing' half the total money now in circulation. If anonymous electronic fund transfers take hold, there may well be a proliferation of players who can, in effect, print their own money. As *The Economist* has pointed out, 'the temptation to move away from a fully backed digital money [may] prove irresistible. Instinct argues that people will want virtual credit, and that it must therefore find a price ... private issuers *that were properly regulated, or enjoyed the absolute confidence of the market*, might prove to have a better "name" than many governments.'[25] In this connection, it is interesting to recall that, before America's creation of the Federal Reserve System in the nineteenth century, some banks printed their own bank notes, although these were often regarded with suspicion and widely discounted.

will then further erode. We will enter a world that is 'comprised of several parallel – and competing – electronic economies,' says Joel Kurtzman.[23] Each will make its own rules. One pundit recently quipped that instead of dollar bills we will also have the choice to use '*Bill* dollars' (a reference to Microsoft supremo Bill Gates). US president Clinton's technology adviser Mike Nelson warns of 'the very real prospect of private international currency ... [that] will affect the ability of countries to manage their own money supply and their own economies.'[24]

The very essence of money – which is needed both as a basis for exchange and as a vessel for savings – is even now being redefined by cybernetic techniques. Says Ernie Brickell, a software wunderkind at Bankers' Trust, 'If numbers are money, then doubling the numbers will double the fun.'[26] But if this e-cash is allowed to take hold in a form that bypasses the networks of remaining regulatory accountability – those currently imposed by nation-states – this will further heighten the danger of systemic collapse. Such costly debacles as America's 1980s savings and loan scandal, or the 1995 collapse of Barings, the venerable British merchant bank – both the result of the underregulated speculative excesses in the derivatives arcade – show how frequently the cost of short-term private speculation is borne over time by the public at large.†

As electronic impulses scuttle anonymously along their gossamer webs in an ever-accelerating inflationary race to maintain their nominal worth, a number of important technical and ethical issues remain unresolved. Who will actually

†For an entertaining sketch of money's mutations through history, see Chapter 2 of J. K. Galbraith's classic, *Money: Whence it Came, Where it Went* (Boston, Mass.: Houghton Mifflin Co., 1975).

control the money supply? How will social and commercial priorities be balanced? On what basis will the value of digital cash be set? Ultimately, what authority will be responsible for backing its worth in case of default? Then there are those troublesome but inescapable errands like detecting criminal money laundering and making sure that taxes are paid. What role – if any – will citizens play in the determination of all these priorities? *The Economist* writes that electronic cash reiterates old questions about money itself 'in a pure almost conceptual form: electronic money promises no intrinsic value, and barely even a trace of physical existence. The Internet is about to push to the limit the question of what makes money worth what it is deemed to be worth.'[27]

It is popular to speak of world finance as a system that is now 'beyond control,' not as one that reflects specific individual, commercial, and political choices. Most leaders, wherever they may be, now feel obliged to trim their sails according to the market winds: decisionmaking is influenced as much by traders' expectations as by local priorities or economic fundamentals. As a result, however, an unrooted financial system is being born with neither the ethical nor the regulatory foundations to ensure its overarching stability. Technology, first brought into the banking business for the sole purpose of speeding up long-distance money transfers, has created the preconditions for financial havoc.

Residents of the developed world rarely look over their shoulders at the ruins of the former Eastern Bloc for intimations of what their own future may bring. But consider: upon their declaration of independence from Moscow, leaders in the Ukraine printed their own currency and styled it 'the karbovanet.' The word – one of the oldest in their language for money – originally referred to precious-metal beads worn by women as decorative necklaces. But, since the new karbovanet was decoupled from any underlying economic reality, it was quickly discounted to the point of near-worthless-ness. In 1994, authorities had little choice but to pulp these colorful charms. It is one of those odd quirks of poetic justice that the banknotes were subsequently transformed into toilet paper – for which Ukrainians could at least find some tangible use. The present trendline in finance, if unaddressed, suggests that future holders of cybernetic dollars may be deprived of even this most elemental of real-world consolations.

Perhaps this sufficiently explains Joel Kurtzman's reluctant conclusion that 'In a vast, free-form, centerless megabyte economy, deregulation may be the opposite of what is needed.'[28]

Sometimes the pressures of the real world can grow too intense and it's nice to pop off for a little break. Let's fly off for a spot of holidaymaking on some Caribbean isle – say Curaçao, in the Netherlands Antilles. Now, you may not realize it, driving along the dusty roads past dozens of nondescript offices, but these unprepossessing woodframe buildings actually conceal the *real* business that makes this island tick. Curaçao is one of the world's emerging data havens.

If you were a cybersleuth, you would quickly understand why. See that building? Climb to the veranda that runs along the length of its second floor, slip quietly past the lawyer's office and the accountancy firm, and you will fetch up at a third door marked 'Entropy Ltd.' In the air-conditioned stillness inside, you will find little more than a half dozen computers, their modem lights blinking away. There is an empty desk. An unoccupied chair. A calendar on the wall, months out of date. This business has been programmed for remote control. You're as likely to find its well-tanned and ever-beaming owner in Miami as at the other end of the island. This urbane Colombian gentleman, who has registered his company in Shanghai, will probably be propped up at the bar. Or else reclining in a vast hammock under the clicking palm fronds at the edge of an exquisite aquamarine sea – toying with his laptop and mobile phone. Occasionally he bestirs himself to attend a video conference, taking care that his computers 'create' a synthetic backdrop that disguises his actual location from the conferees. But you will be hard put to establish that Entropy Ltd is a virtual casino. It is 'based' in the minute pulses of positively and negatively charged energy that whir through the electrosphere, log on gamblers from around the world, and suck in their electronic cash at every 'turn' of the virtual wheel. The profits are automatically transferred to places like Anguilla and Liechtenstein, where they join other untaxed 'nominee accounts' in which most of the world's private wealth has by now been salted away.* Every day, legions of hope-

ful gamblers fire up their modems for a shot at some fantasy-like windfall. Now and again a sizable bounty is actually paid, though there is no way to confirm that the posted odds are actually maintained. No one has any idea where this casino is actually located; no one knows to what extent the tables are honest or rigged. To the gamblers, the casino is a mere 'interface' on their screen: all the distracting bright lights and magic

*A single example: the Caymans, a set of three low-lying islands under British jurisdiction about 180 miles to the northwest of Jamaica, have a population of about 30,000. The colony also has substantially more registered companies than actual citizens, and a profusion of banks that reported deposits totaling $460 billion in 1995 – virtually all foreign assets attracted by the islands' tax-friendly clime. (*Money Laundering and the International Financial System*, IMF working paper, May 1996, cited in Stephanie Flanders, 'Cleaning up the global economy,' *Financial Times*, 28 August 1996.)

of 'The Desert Flower Gaming Den.' It is a casino that pays no tax and answers to no one.

What an appropriate spot to ponder some of the disturbing contrasts between the regulation of trade in physical versus virtual space. It is miles removed from that windswept quay on the east coast of England where, on a real-world March day in 1990, a huddle of rain-lashed officers of HM Customs & Excise gather to announce that they have uncovered a remarkable cache of contraband. Dangerous chemicals or narcotics, you presume? Think again. A set of eight bizarre steel *tubes* – rifled on the inside – comprising a massive 'Supergun' barrel that is destined for an ominous Iraqi weapons project code-named Project Babylon. The tubing is stacked about with a misleading label that reads, 'Goods: Metal Parts: Oil Pipes.' Anyone can see they are intended for something far more sinister than pumping crude. The long-range Iraqi Supergun had been painstakingly machined by Sheffield Forgemasters in the UK after several years of clandestine design, planning, and production, all coordinated by intelligence agents working under the direction of Saddam Hussein. It was on the verge of being released into a region that was rumbling like a volcano set to explode. Had it arrived, Saddam could have delivered nuclear warheads directly into Israel and other Middle Eastern states. The full story behind the discovery of this gun will probably never be told. It is swathed in layers of skulduggery and subterfuge – part of a vast arms-trading network that operated with official collusion of intelligence agencies in Washington DC, Bonn, London, Paris, and Rome (just to name a few) as Western agencies encouraged such sales for a time in hopes of weaning Saddam away from his client relationship with the former Soviet state.

Fortunately, Saddam Hussein's Supergun was never sent on its way, but it casts an uncommon light into the real nature of the digital world toward which we are slipstreaming so fast. With the devil that we know – reassuringly *physical* environments – cargo ships slip into berths every day. They unload containers onto the quay. These are tallied against manifests. They can be subjected, where necessary, to tax or seizure. But, in the coded interstices of cyberspace, value is contained in the ultimately mobile, often untaxable, and potentially untraceable form of binary code. Such code might correspond to a tax-allergic fortune (although most of these are already thought to 'reside' in existing financial havens and offshore accounts) or to information on private individuals illegally compiled in offshore data havens (a number of which, as noted, have already sprung up).* As e-cash starts to take hold, says Wim Duisenberg, the Dutch-born chief of the European Monetary Institute in Frankfurt, who has been widely tipped to head the future European central bank, 'there is a definite concern about a potential loss of transparency in the financial system as a whole.'[29]

*One example is Black-Net, on which any and all kinds of information can be bought and sold anonymously. See Tim May, 'Crypto anarchy and virtual communities,' *Internet Security*, April 1995, pp. 4–12.

As the definition of what constitutes information widens, and cybernetic environments become the central focus of commercial activity, elected authorities will discover that *monitoring* the shape and direction of the economic game will become increasingly problematic. 'Things aren't always visible in the virtual environment,' remarks Dorothy Denning. She pauses to let the implication sink in: that data can be encrypted (or secretly coded) in such a way that not only is it impossible to read, it is also impossible to identify from whom or where it originated. Denning is an expert in advanced cryptography, held in high esteem by colleagues at home and abroad. As she sees things, 'If you're a major company, you will almost surely use crypto to prevent your competitors from accessing your trade secrets.' This is a powerful weapon that can be turned against public policymakers as well. 'The widespread availability of unbreakable encryption coupled with anonymous services could lead to a situation where practically all ... electronic transactions are beyond the reach of any government regulation or oversight. The consequences of this to public safety and social and economic stability could be devastating.'[30]

By the turn of this century, analysts believe that about $600 billion worth of goods and services will be traded over electronic networks. But even today, observes one London-based communications industry executive, 'there are probably hundreds of millions of dollars worth of intellectual property crisscrossing national borders ... without paying customs duties or taxes. This is happening not just on the Internet, but on thousands of private and commercial networks all over the world. We have not yet invented the electronic customs inspector for electronic objects moving at the speed of light over the invisible superhighways.'[31] In other words, even though more wealth is being generated by the system, a growing proportion of it is being shifted beyond the regulatory reach.

According to Tim May – a self-professed 'anarcho-capitalist' who has frequently crossed swords with Dr Denning both on the Net and in print – advanced encryption technology is 'altering the conventional "relationship topology" of the world, allowing diverse interactions without external government regulation, taxation, or interference.' May foresees a day when encryption tools will make both money and the details of everyone's personal wealth inaccessible to snooping regulators. Eventually, this will pull the plug on government efforts to collect tax and thus, crucially, to finance its activities. He regards this quicksilver flow of information, and the bypassing of national jurisdictions, to be a legitimate form of 'regulatory arbitrage.' He is firmly persuaded that this 'is liberating individuals from [government] coercion.'[32] If May is right, the trend is toward a multi-tiered system of public responsibility in which the largest and technically strongest will contribute, proportionally, the least. In fact this is already taking place. Governments feel compelled to offer large companies 'fiscal incentives' – such as substantial cuts in tax liability – in order to persuade them to make job-producing investments in specific locales. The companies, being more mobile, are in a strong position to play one offer against the next to obtain the most advantageous deal. As a matter of fact, according to recent World Bank data, some two-thirds of all major investment decisions in 1993 were based on the availability of tax and other inducements. According to the UN, however, 'incentive competition between governments is very costly ... and can generate inefficient investments with disappointing results. If it goes too far, not even the "winning" country obtains a net benefit.'[33] In short, within the distributed systems of the digital world as currently designed, only the fastest and most efficient 'nodes' prevail, yet their victory will prove short-lived at best.

The immense forces tearing at the already filmy fabric of postmodern society as it is 'reconfigured' into a cybernetically driven market system are not being factored into any sensible vision of future development.[34] Citizens face a radical devaluation in their hopes and dreams. It is hardly a coincidence that the world on the cusp of the twenty-first century finds itself fractured by a radical increase in regional, ethnic, and fundamentalist conflicts. These are fueled by people's desire, however irrationally expressed at times, to recapture a more reasonable degree of self-determination. This is true even in Europe, where there remain pockets of visceral hostility to further European integration. In numerous other countries – including Algeria, Turkey, the former Soviet Union, China, and even the United States itself – government has become a target of the ire that should more rightly be directed at specific politicians and an unaccountable financial elite.

Now, as ever before, it is only appropriate levels of common sense and moderation, only workable networks of accountability and ethical restraint, that can form the basis for the kind of trust on which long-term social and market stability depend.[35] The economist Friedrich von Hayek began his influential book *The*

Fatal Conceit by quoting a colleague to the effect that 'freedom is not, as the origin of the name may seem to imply, an exemption from all restraints, but rather the most effectual application of every *just* restraint to all members of a free society whether they be magistrates or subjects.'[36] Oliver Wendell Holmes, the distinguished American jurist, observed that taxes may be painful but they are also 'the price we pay for a civilized society.'[37] Likewise, the search for a better world requires that individuals within a society share costs and sometimes accept lower short term profits in pursuit of an equitably shared and sustainable prosperity. For instance, it is trivially easy to use one's personal investment account as an instrument with which proactively to reward enterprises that operate in more responsible ways. Micro-banking and ethical and eco-friendly investment strategies are also positive steps toward a brighter future. There are others. All of these low-tech expedients require that the well-to-do – those who actually *have* surplus resources to invest – feel a sense of commonality as social beings with their less privileged compatriots. For the wealthier, the task of keeping well-informed, and identifying the intimate relation between (for instance) a tactical stock purchase and its long-term effects, will require concentrated effort in this more complex and digitally distributed world. So will finding the self-discipline to confront the resulting insights and act upon that knowledge accordingly. But, by forging one link in a chain of mutual accountability, anyone can help ensure that finance is reintegrated into its wider constituency of life. Thus an 'uncontrollable' market can be encouraged to serve the creative purpose that is the only legitimate justification for its existence.

Yet, as a more fully networked world takes hold, the costs of maintaining civil society are bumping up against the discount doctrines of anarchic individualism. The essential argument is that technology, coupled with a libertarian 'free market,' should be accepted as a public good in its own right. The techno-libertarian future, as outlined by the most optimistic among the digital elite, seems to offer glittering promise without associated responsibilities or costs. The message is that you can get something for nothing, and people should always treat such an idea as intrinsically suspect.

In fact an absence of democratic oversight will merely give rise to a situation in which *the strong* are increasingly free to transfer the costs of their lifestyle to the weak by technical means. Is this acceptable? Do we want to live in such a world? Future governments – lacking the power sufficiently to tax or regulate worldwide movements of the means of exchange and savings – will have no choice other than to seek a proportionally greater amount of their revenue from sources that are physically rooted in place.

As Columbia University's Eli Noam inquires, 'What are you going to do if they slap a high tax on food? Eat less? Or grow your own vegetables? Very possibly. But one thing's for certain: you're sure as hell not flying to Tokyo for lunch.'[38]

6

Culture versus System: Landscapes

Cyberspace. A consensual hallucination experienced daily by billions of legitimate operators, in every nation ... A graphic representation of data abstracted from the banks of every computer in the human system. Unthinkable complexity. Lines of light ranged in the nonspace of the mind, clusters and constellations of data. Like city lights, receding ...

WILLIAM GIBSON[1]

It is night and the blizzard has set in, sending grains of glacial snow scudding through a crack under the oak door. The month is January, and, high in the Austrian Tyrol, I am enduring the unexpected adversities of my first alpine ski tour.

These excursions teach you to cultivate a relationship with heights that is quite different from that which you absorb when downhill skiing. For one thing, there are no lifts to whisk you to the peaks. If you want to reach the high passes, then you simply have to stretch snow-gripping skins across the base of your skis and pole your own way up. The climbing is hard. Once you reach the high snowfields, of course, you feel an exhilaration that has been heightened by that work. Suddenly you find yourself in a world apart, gliding from one alpine hut to the next. Left far behind are all the plastic encrustations of a modern commercial resort: the overloaded tour buses, the trendy boutiques, the staccato percussions of rap mixed with dull disco beat. Instead, you have the calm majesty of the mountains themselves. Climbing, you are surrounded by a pure and airy silence, broken now and again by the whispered beat of crows' wings.

Exposed in this windswept domain, tethered by a slender rope behind your companions as the stormclouds whirl and gather overhead, you ascend the vast, glistening slopes, step by measured step, nearing an invisible spot under the next peak. There, anchored improbably in a sheltered lee by the high pass, a weathered gray stone chalet promises welcome, shelter, and wood-paneled warmth. When night falls, you find yourself crowded around a common table, sharing an evening meal. Faces glow in the gas lamplight. Winds gather force beyond the door.

As the plates are cleared, three climbers unsnap their small and battered instrument cases. They begin a song. It is an old song of the Alps, and it has been long-since forgotten in the developed valleys 10,000 feet below. With the right ears, you can almost *hear* the physical and cultural contours of the mountains in such airs. The melodies rise and fall with the same breathtaking exuberance as the jagged peaks. The lyrics have the same abruptness, the same undertone of awe, and an awareness of mortality that comes with life truly lived at the edge. In the commanding immensity of these precarious spaces, the singers imply, it takes balance and care to achieve sustained progress of any kind.

The song draws to an end and there is laughter and backslapping all around. A bottle of '*Bergwasser*' is tilted and recorked. As the wind sweeps past the eaves, I have a keen sense that there are ethical values only landscape can convey: a universal relationship between people and the earth that runs far deeper and is more spiritually binding than shallow and divisive nationalism could ever allow.

Mountain people the world over have evolved singular adaptations to the unforgiving landscape of their lives. They express these accommodations in a unique vocabulary, not only of words but of gestures, practices, and implicit beliefs. A mountain-dweller in the south of Germany might use the same literal language as a shipowner on the North Sea, but he uses it to lay stress on different things. Significance is contextual. Ask an Inuit tribeswoman of arctic Canada about the importance of water and she'll give you a radically different answer than a tradesman of the Arabian sands.

The flood has a different symbolism to a farmer on the Indian sub-continent, anxious for the blessing of monsoon, than it has for a Dutchman intent on keeping the North Sea at bay. What's 'liberty'? It depends on whether you're asking a silk-suited Russian currency trader or a Burmese political prisoner newly released from house arrest. And a 'sensible' pace? A Caribbean bank clerk, floating through the years of unremitting tropical heat, might think she's offering gracious service to the bad-tempered tourist just arrived from frigid Stockholm for a hasty week's break.

Literatures, economies, political systems, and entire world-views take shape in singular, unrepeatable intersections of time and place. Any time we refer to a person's being Japanese, Egyptian, or Miskito Indian we are using a kind of shorthand for complex states of mind, special patterns and perspectives, distinct priorities that have evolved and endured because of the way these people have embraced the limits and utilized the unique possibilities of their given space. Because they have been obliged to coexist, both with the landscape and with each other, they have evolved special ethical frameworks. They have learned to season their aspirations with the salt of realism; they have come to realize that Nature, in all her bounty, will ultimately spurn any lover who is consistently untrue. From the earliest times until now, creatures have always been obliged to test their collective models of reality against the demands and disciplines of the physical world.

If you visit a paleolithic cave site in France or Spain, you can see the magical drawings which crowd the walls and remind you of an oft-forgotten fact: namely, that every culture's world-view and every individual's personal consciousness are nothing but simula-tions, living dreams, drawn from the materials at hand. The cave drawings also highlight the extreme reverence and sensitivity of those ancient dreamers to the surrounding patterns of actual life. These hunters took pains to study the habits of their prey (which in those warmer days included rhinoceros and woolly mammoth) and to cull them in numbers that would be replenished year after year. When abroad on a hunt, they also knew to differentiate the out-lines of stalking lion (then plentiful) from the tumult of tawny grass. And they learned to read the correct meanings in the movement of clouds and the wind. In cybernetic terms, they were amplifying the essential *signals* and suppressing the surrounding *noise*.

The choice of what to select as signal and what to reject as noise is just as momentous today, after centuries of evolution, even if the ambient realities are radically changed. The residual skills of a hunter – notably the instinct to gather and hoard – are poorly adapted to the contemporary challenges of escalating overpopulation, resource scarcity, and sociopolitical ferment. So, while the basic stakes of survival remain unchanged, the drive to find a more sustainable order has leapfrogged from a local to a global concern. The underlying paradigm of materialism is bumping up against physical limits in the basic carrying-capacity of the earth. Where, earlier, the drives to compete and survive were one and the same, now they are frequently at odds: repeated individual acts in pursuit of immediate material advantage can carry devastating collective long-term costs. A period of postwar superabundance must come to an end. Can the developed West accept a profound change in its high expectations even as the rest of the world struggles to catch up? Can the splintering human tribes find a vocabulary which admits to mutual accountability for the future of the earth? From one human age to the next, there are a thousand defining moments, great and small. People are continually confronted with the imperative of choice. Whether by commission or omission, they amplify certain basic priorities at the expense of others, and the decisions are subjective, binding, and frequently inescapable. *Choice* – as the Spanish philosopher Ortega y Gasset rightly proposed – is 'the defining act of culture.'

Singular and subtle choices have been invisibly nested in every artifact we've ever produced – from songs to political systems to consumer goods and services. Just picture a cabinet of curiosities. Imagine that each shelf has just one single item on display – a small handful of Japanese rice; a rough ingot of Sheffield steel; a slab of Breton butter; a bottle of Bavarian beer. Hidden within the superficial shape of each of these are specific uses of the land and its resources, unique culinary traditions, and a singular set of social interrelations. The self-styled car buff will grip your arm with excitement after a spin at the wheel of a good British or Italian sports car, insisting that these machines exhibit an altogether different 'road spirit' than those produced by the Koreans or Japanese. Any rank amateur can admire the unmistakable *Frenchness* of the classic Citroën – a plush and saucy affair that virtually levitates on a bizarre system of pneumatic pumps.

But the natural environment has largely ceased to be a conservatory of culture and the arbiter of ethical outlooks: with the advance of economic and technical development, men have confidently imprinted their own choices and ambitions upon nature instead.*
Western culture is one that is psychologically and spiritually distanced from the landscape. Nowadays, nature is either romanticized (as an Edenic garden) or objectivized (as an inexhaustible source of material gain).† Life has already been decoupled from the rhythms of the physical world: the trading-desks of world finance glow with bright pulses of light night and day. Previously localized economies are being more intimately enmeshed in

*Simon Schama, the British writer and historian, remarks how 'many of our modern concerns – empire, nation, freedom, enterprise, and dictatorship – have invoked topography to give ... ruling ideas a natural form.' The bison became an icon to the spirit of Polish nationalism; the bald eagle was transformed by Americans into a symbol of independence; they carved effigies of their leaders into Mount Rushmore rocks revered by the Lakota Sioux. (Simon Schama, *Landscape and Memory* (New York: Alfred A. Knopf, 1995), pp. 10, 17.)

†For most urban dwellers, the chastening engagement with nature is a thing of romantic dreams: the natural world has become a 'garden, or a view framed by a window, or an arena of freedom' as the novelist John Berger remarks. He goes on to say that 'peasants, sailors, nomads have known better. Nature is energy and struggle. It is what exists without any promise. If it can be thought of by man as an arena, a setting, it has to be thought of as one which lends itself as much to evil as to good. Its energy is fearsomely indifferent. The first necessity of life is shelter. Shelter against nature. The first prayer is for protection. The first sign of life is pain.' Beauty and transcendent values are rendered all the more poignant, he notes, because they exist *in spite* of these harsh adversities. (John Berger, 'The white bird,' *The Sense of Sight* (New York: Vintage Books, 1993), p. 7.) Nevertheless, in sharp contrast to the societies of Asia and the Americas, Europe evolved a culture of technology based on 'the idea that human achievement and material betterment were to be won by *opposing* nature,' according to Kirkpatrick Sale. 'Nowhere else was the essential reverence for nature seriously challenged.' Europe, as long ago as the fifteenth century, was already suffering from the effects of its depredations on nature in the form of overcrowding, plagues, and scarcities on an unprecedented scale. At just about that time, co-incidentally, it invented gunpowder and took to the seas. (Kirkpatrick Sale, *The Conquest of Paradise* (London: Hodder & Stoughton, 1991), p. 88.)

this throbbing, globally interdependent web. The flywheels of 'progress' are unsettling regional relativities and casting us onto a broader and more cosmopolitan sea. 'Lifestyle,' once an outgrowth of the landscape, has evolved into a *product*. In the process, the range of our possible choices is being radically redefined. Our professional dreammakers have perfected a form of mass persuasion which is called advertising – and which is now broadcast on a transnational scale. We live in a world of manufactured expectations – one in which televised images, streaking from the urban boardroom to the impoverished *barrio* in their fiery millisecond bursts, illuminate the entire 'global village' in a flickering, transient light.

The paramount human experience, writes the author Salman Rushdie, is increasingly one of relentless psychic migration, of unsettled movement through an ever-shifting *mental* terrain. Identity is now a personal *construct*, rather than something shaped within the context of a relatively stable social space. The whole of life becomes a discontinuous collage. We experience 'multiple ways of being.' We become cultural consumers, capable of constantly defining and redefining ourselves, shifting our physical and psychic venues, choosing from the ever-widening array of options at hand. It is also striking that this freedom seems to have opened up what Rushdie calls 'a large God-shaped hole,' a spiritual vacuum that coincides with the return to cultural chaos, a volatile condition in which reality, 'like any other artifact, can be made well or badly, and ... can be unmade.'[2]

Marshall McLuhan, the prophet of cybernetic society, reflected long and hard on the cultural impact of a more visually oriented and electronically mediated lifestyle. He concluded that its main effect would be to unify diverse cultures and weave the family of man into a single tribe. Decades later, his global village is arriving in an unexpectedly disturbing guise; East and West, the landscapes of inherited identity are being swept by powerful homogenizing tides. The very speed and discontinuity of this transformation have triggered a reflexive and fragmenting search for extreme certainties and fundamental truths. Thus the throb of the tribal drum, so eloquently celebrated by McLuhan, is actually proving to be one of the most unsettling reverberations of our day. The US historian Samuel Huntington is convinced that we are moving into a period that will be marked by a 'clash of civilizations,' a swirling maelstrom in which the supposedly unified worlds of Confucianism and Islam will confront a universal 'ethic' of free-market capitalism and the secular gospel of consumptive growth.[3]

But a circle has invisibly turned. Five hundred years ago Europe first set out on its adventure of outward conquest. It was driven by an outlook 'in which rootlessness and restlessness became adventure and curiosity, in which there was little room for constraints and limits and restrictions, in either the physical or intellectual [domains].'[4] Now, every nook of the world's physical territory has been minutely explored and spoken for. Many of the former colonies are beating the aggressor at his own game. An awareness of impending limits is once again pressing in. And so, crouched in the dawning light of the 'information age,' a mainly male and mainly Western technologist is devising his most audacious leap yet. He is devising an *alternative* landscape, a digitally mapped electronic frontier into which he can project his boundless and unquenchable dreams. 'Like Shangri-la ... like every story ever told or sung, a mental geography of sorts has existed in the living mind of every culture, a collective memory or hallucination, an agreed upon territory of mythical figures, symbols, rules, and truths,' remarks Michael Benedikt, an architect specializing in the topology of this emerging world. He is galvanized by the fact 'that technologically advanced cultures – such as those of Japan, Western Europe, and North America – stand at the threshold of making that ancient space ... uniquely visible.'[5]

So, while Samuel Huntington looks ahead and sees a looming clash of civilizations, the future may actually involve a more fundamental 'war of worlds': a struggle between real and virtual reality.

With their clanking metal rods and smoke-belching fire-tube steam boilers, the first Victorian passenger trains made an immediate and unmistakable impression both on the landscape and on those who rode on them. For the brave and intrepid travelers, hurtling toward their destinations with unaccustomed speed, there were some disquieting shifts of perspective in store. Gone was the stately progression of the horse-drawn carriage, with the landscape flowing past as a seamless whole. To those flying down the rails, looking this way and that, one sight could seem sequentially unrelated to the next: suddenly they were mentally *assembling* their world-view out of the blast of images assaulting their eyes. One's sense of passage through space was fragmented and discontinuous, and the sudden perceptual chaos persisted until these new, rapid-fire shifts could be internalized and absorbed. As if this were not enough, one's relationship with the countryside also profoundly changed. Walking or riding on horseback, one felt the physical contours of the earth, if only because, upon reaching a hilltop, one felt the organic need to stop and cool off before moving on. When passing through the country in this way, one was also well and truly *connected* with it; one noticed the wildflowers growing around the roadside inn.

A process that began with locomotives, gained momentum with the telegraph, and took on a dizzying spin with the development of private cars, passenger airline travel, commercial radio, films, and TV, brought blistering mobility and new forms of 'connectivity.' It brought unprecedented states of individual and existential freedom and laid the foundations for industrial economies built around quickening life cycles and new trajectories. The telegraph 'freed' information from the limited speed at which people and beasts could travel to deliver it. Information first flew across the plains on electric wings – then it was committed to print. The unexpected consequence was that news became detached from all the local conditions and contexts that made it meaningful. A connection, once experienced directly, was now vicariously consumed via an intermediary electric web. This was great news for business. Once information could be transported by electrical means, it was only a short step before such data became a market commodity.[6] Little wonder that the great newspaper empires soon followed: local gossip became a global sensation.

One of the most profound trends over the past few centuries has been precisely this disengagement from local contexts in favor of the cosmopolitan benefits offered by enhanced mobility. It may even be possible to think in terms of an Age of Mobility: a period in human history when relatively stable, ubiquitously shared, and living presences, like the natural landscape, were exchanged for discrete, individualized, and fast-moving enclosures, like cars and virtual software 'environments.'

In any case, before the Industrial Revolution people's lives were largely encompassed by the strictures and structures of physical happenstance. Harvests were bountiful or scant. Storms and revolutions swept the earth. For better or worse, reality was inescapably shared. This, in turn, conditioned the rituals and codes by which people lived. The larger sweep of human history has been characterized by this authentic connection with the earth – one that inspired an animistic view that all of terrestrial life is infused with a spirit and should be respected accordingly. Of course, contemporary science scorns this perspective, but it was nevertheless creative in its own way. How? It set up a fundamentally ethical tie between individuals and the network of life in which they take part. The notion that the world could be dominated by sheer force of will – or indeed owned – would have been rejected as dysfunctional and absurd. One was *accountable* to one's environment, both natural and social; it had to be cultivated with a long-term view. Life's abundance, when it came, was accepted with modesty and thanksgiving; it was seen as a gift, whatever one's strivings, and not merely a reward for personal achievement.

Today, in contrast, for any sufficiently wired Westerner residing in a community where the soil has been exhaustively overfarmed, where the social codes seem either too stringent or too relaxed, or where one is confronted with awkward social situations or unexpected developments, a fresh landscape is just a mouse click away. This contemporary explosion, fragmentation, and proliferation in the *landscapes* of experience, and in the possible trajectories of personal escape, is unique. Even the railway networks were built in physical space; the tracks rationally connected cities with seacoasts and industrial railheads. Passengers and freight were picked up and dropped off along a network of fixed points. People traveled according to the same fixed timetables. Now the windows onto life are multiplying. In an alternative – simulated – space, the avenues of mobility are shifted out of nature, through the screen, into the electronic Net.

The network's topology is far more radically discontinuous and dispersed than that of the natural world. It is characterized by multiple timelines, a fragmented profusion of chaotic images, and a galaxy of unconcentrated and ever-shifting electronic 'nodes' and 'sites.' People can 'interact' over vast distances at low cost, and thus project their personalities and priorities into far corners that they could never physically reach. Moreover, whereas mechanical inventions enabled people to *leverage* their limited power and to pursue their goals at *arm's length*, networking technologies altogether eliminate these human or natural references and substitute those of placeless abstraction. Whether the sun is up or down, whether the surrounding culture is Hindu or Latin, the world economy will tirelessly whir on, like the computer it has become, regardless of physical time and place – provided, needless to say, that there is electricity to power it.

What sort of impact will this have on a culture's life and its collective dream? Individuals can now 'visit sites' and jack into spaces that satisfy some shifting urge – then instantly jack out if confronted with unwelcome demands. Grabbing frames and moving on, one's identity becomes a collection of elective affinities – of choice from a menu of options – an agglomeration of individualities ultimately built on unmodulated *desire*. Indeed, what makes cyberspace unique is this expanded freedom with which we can project ourselves between 'freely chosen' environments, in effect inventing and reinventing ourselves at every click. We become less concerned about the expectations and responsibilities that prevail in static communities: our possibilities are bounded only by the number of nodes into which we can fluidly shift. As the process escalates, it becomes easier to gradually opt out from the messy but essential business of accommodating difference and bridging gaps in the wholeness of shared space – where our bodies (if not our minds) are still inescapably situated. For the first time in human history, humankind is trying to evolve without the disciplines and inspirations of physical space. And, in the process, those systems we once knew as human culture, which were patterned and conditioned by earthbound circumstance, may begin to break up and then coalesce around powerful new hubs of digital *signal management*. These are being built to satisfy the primal needs and unique desires associated with passage through a virtual landscape.

The world, as it approaches a new millennium, is clearly operating in a more accelerated, reactive, and rapid-fire environment than ever before in human history. Interestingly, the emerging electronic landscape has been insinuated upon the scene in gradual and subtle stages, each seemingly imperceptible but each nevertheless reinforcing the next. For example, footloose electronic cash quietly reinforces the rise of *electronic* spheres of commercial power, which in turn has reduced the accountability of private enterprise to elected governments. (See Chapter 8 for more detail.) Likewise, simple mobile phones are reinforcing the capabilities of the portable PC and, in the process, reconfiguring human patterns of private life and work. (Whether attempting to relax at home or traveling busily on the road, for instance, the contemporary executive is more and more tethered to the Net. If he or she tries to 'log off,' this jeopardizes continued employment and economic survival.) This tightly integrated yet systemically fragmented pattern of cybernetic life only began to crystallize with the comparatively recent spread of its enabling technologies.

In the late 1960s there was just one submerged telephone cable that spanned the Atlantic Ocean. On a good day it could handle a maximum of eighty-nine calls at a given time. A three-minute phone conversation cost you the equivalent of $100 today. And if you wanted to talk with somebody on the other side of the Pacific you were offered a glitchy radio link that bounced your voice off the earth's lower troposphere.[7] It's hardly a big surprise that, for most people, long-distance telephony occupied a fairly isolated island in the broad sea of everyday life. In fact, apart from a few bankers, well-placed businessmen, and high policymakers, most people's lives sailed along quite smoothly without even nearing the shores of what we now call the Net. After all, people have traded goods across frontiers for millennia. And in many ways, business was just as 'global' two hundred years ago as it is today.* (Indeed, statistically, the flow of capital among trading nations, relative to their total GDP, was *higher* in the nineteenth century. Migration across borders was easier as well.) What's really changed is the *context* in which business is managed and regulated: it now operates within a digital landscape. There is also a different regulatory regime and a different overall pace. (Crossing the Atlantic, until the 1950s, involved time-consuming travel by boat. There were no fax machines or cellular phones, and the notion of computer networking seemed like a contradiction in terms.)

Three intervening decades have radically reconfigured this scene. Business now *resides* on the Net, and a growing number of people now live digitally intermediated lives. Instead of driving to work, for example, work comes to them via data bits over the modem and the mobile phone. Meanwhile, that transatlantic phone cable of the 1960s has evolved into a hydra-headed tangle of high-capacity fiber-optic filaments, satellite networks, and microwave facilities capable of simultaneously coordinating over a million phone calls with a torrent of pictures and text. All of these data are expressed in a language of digital bits: the sheer volume is beyond comprehension. Virtually every sinew, muscle, and ligament of the body electric is animated by this binary beat. We swim in the sea of data that floats invisibly under the buoyant surface of postmodern life. A day hardly passes without enthusiastic paeans to an emerging 'world culture' that is supposedly taking shape: 'Apart from differences of skin and face and language, continents could as well be different floors within . . . a unitary

*The current process, often described as a *globalization* of big business, is in fact its *digitization and electronic networking*. Networking coincides with and drives the trend toward corporate concentration, which is aimed at attaining economies of scale and scope. Whereas in the early 1980s almost every developed country had a handful of radio producers, car makers, and textile manufacturers, all of which had their counterparts in nationally based unions and regulatory regimes, most capital-intensive industries are now engaged in a vast reorganization within a unitary information space. Only a handful of cybernetic giants will remain in their respective business sectors after the smoke has cleared. At present, such enterprises – sometimes called transnational companies (or TNCs) – control about a quarter of world GNP. It remains to be seen how this proportion will increase.

urban supermarket.'[8] To be wired is to be 'freed from time and space.' To listen to the spin doctors is to conjure a future that suggests nothing so much as an endless (and highly personal) out-of-body experience.

For those who can stand back from this hubbub, contemporary life starts to look like a party thrown into a frantic pitch by its subliminal awareness that the hour is late. Some grasp for the dregs of booze that remain. Others unsteadily wonder whether they, or anyone else, are still sober enough to find their way home. Maybe the invention of cyberspace can actually be understood as a subconscious collective response to a growing apprehension about physical limits – about the need finally to confront the cold light of choice and inescapable sacrifice – and reflects a wish to create a carnival-like arena where such restraints will not obtain. It surely takes a powerful Zeitgeist to print a million sweatshirts with a slogan like 'There *are* no limits' and slap them onto the backs of so many pot-bellied and jogging big-city financiers. In London, one manufacturer of satellite telephones takes out a triumphant newspaper ad that invites us all to 'Get *wired* and get *free*!'

The exquisite irony, of course, is that, the more 'wired' we get, the more dependent we grow on the intermediating technical web. Despite the steady hype about how networked virtual worlds will 'render the mental landscape of our electronic culture uniquely visible,'[9] the essential point about the Net is precisely its growing *invisibility*. Cyberspace is comprised of far more than the much-hyped Internet or new virtual worlds. It is the total communications space that saturates our lives. The Vancouver-based author who coined the term, William Gibson, has famously described cyberspace as 'a *consensual hallucination*.'[10] Of course, there is nothing hallucinatory about the physical armature of digital hardware on which it is built – all those optical fibers, copper cables, electronic circuits, and all the rest – nor about its energizing by electrified code. It all adds up to an existential sprawl that is so vast and ubiquitous as to seem palpably real. This is the *information space* of modern society. It pulses with blasts from the sophisticated production studios, satellite uplinks, digital delivery systems, and receivers of the global media. The electrified walls of the world's cybernetic coliseum echo with torrents of infotainment – a roaring metapresence so powerful as to decimate sustained and concentrated thought. The whole spectacle, shaped by pervasive new signal managers, is infinitely more compelling than many of the responsibilities and challenges of terrestrial life.

Life-experience today is firmly divided between the physical here and now and a 'someplace else' immanent in the glow of audio-visual space.* The virtual world is thus a place in which we already live. The social, economic, and political lifelines of the world are now evolving almost exclusively within it. For example, this is where large corporate alliances busily coordinate their quest to profitably balance 'efficiency,' flexibility, and economies of scale. It is where data about consumer demand, production flows, finance, and all the rest are managed and shaped. This is the place where our savings 'reside' in the form of skittish blips which are used by large banks and investment firms to profoundly reshape the relativities of contemporary work and play. If there is a revolution under way, then it lies in the way all of these networks are being interconnected to form an ambient, enveloping, mind-molding, sensory sphere. Ironically, it also lies in the final disconnection of humankind from the earth. Granted, over half of the planet's population is still producing low-tech crops – and mothers everywhere still give birth to baby boys and girls – but for most of those who reside in the developed world the terms and conditions of life are no longer decided in nature. They are being shaped in the lattice of the Net. Indeed, it

*One AT&T executive writes that 'as more and more time is spent interacting with ... virtual culture through globally driven networks, local culture does ... become diluted or dissolved ... ' due in part to 'an appeal of wealth and power more than that of self-respect and self-fulfillment.' The result is a generation of TV and video-game addicts whose attention is 'focused elsewhere than their immediate surroundings.' Thus, disconnection from nature makes it possible for people to reinvent themselves in hyper-real, or virtual, ways. (Frederick S. Tipson, 'Global communications and the sovereignty of states' (pp. 1–14), an unpublished AT&T discussion paper on 'non-state actors' in world politics, dated April 1995 and circulated among a group of industrialists and high policymakers attending a Council on Foreign Relations forum in Washington DC.)

has become very difficult to escape the cybernetic embrace. The digitization of the natural landscape has set the stage for its ultimate privatization: one that heralds a radical change in how the very essence of our natural patrimony is managed. For, if empires expand along with their networks of communication, as Rome expanded along a path cut by its roads, then we can be said to be embarking on a new phase – one in which the world is dominated by *empires of communication*.

In the early 1990s a twenty-six-year old Guayami tribeswoman checked into a hospital in Panama. She had a blood sample taken while being treated for her ailment. The sample found its way to certain genetic scientists by unknown means. After some time had passed, her entire tribe was astonished to learn that its genetic blueprint (which is distinguished by a unique resistance to particular diseases and is therefore of intense potential value for pharmaceutical development) was suddenly being *patented* by the United States.[11] Similarly, in 1994, the medical group Smith-Kline Beecham paid HBS, a smaller biotech enterprise, $125 million for *exclusive* access to a private database containing the DNA sequences of a collection of human genes; many others, less immediately valuable, are posted daily on the Internet.* Scores of plant scientists working for the major chemical groups are busily synthesizing new medicines and crop strains based on substances naturally found in the world's dwindling natural ecosystems. Ironically, once such crops or remedies are patented, the indigenous tribes using the natural product, extracted from local rainforests by traditional means, are transformed into 'software pirates.' They are criminalized for unknowingly violating a 'property right' claimed by some distant high-tech enterprise – one that works to priorities they neither recognize nor comprehend.[12]

*Where exclusivity is not imposed, the practice of charging royalties on such data also acts as a form of economic censorship because it effectively limits access to those with the means to pay.

There is little doubt that the Human Genome Project, a worldwide biological effort to map and determine the chemical sequence of the 3 billion nucleotide base pairs that comprise the human blueprint, is likely to produce profound scientific insights. It may explain the mechanics of terrible diseases including cancer, cystic fibrosis, and Alzheimer's – and eventually yield possible cures. But the essential choices posed by sequencing the human genome are fundamentally ethical. They concern the very management of life – and thus the way in which humankind perceives both itself and the natural landscape on which it ultimately and inescapably depends. Questions include: How will the difference between a 'normal' and an 'aberrant' human configuration be determined? What will be 'superior' and what will be deemed 'inferior'? What's more, how should we act on that information, and who should decide in each case? There are further questions, such as whether patients can be obliged to seek treatment for certain conditions, and whether insurance companies can demand genetic screening and deny coverage if such conditions subsequently appear. What rules will govern the confidentiality of this data? Looking ahead, moreover, and considering the powerful primal incentives to protect the dominance of one's own kind, we can already foresee a move toward the eugenic engineering of a 'master race' (and, by default, a 'genetic underclass'), just as Huxley warned in his *Brave New World*.* And, as if this were not enough, there is the unnerving possibility that we may begin to regard *ourselves* cybernetically: that is, as a mere collection of interacting genes that 'operate' as systems in complete isolation from the social and natural circumstances in which we live. The perverse result of opening this new informational frontier may well be that the value of a culture's heritage will be reduced to a measurement of its market worth.

*In an article entitled 'Religious leaders oppose patenting life,' released on 18 May 1995, the *Los Angeles Times* News Service reported on an interdenominational appeal in which 'The religious leaders noted that research on human embryos is permitted, and the next step will be the modification of these embryos to create "designer human beings." At first, the stated objective will be to correct serious genetic flaws that would lead to disease or disability, but the same techniques might be used to endow the human embryo with physical and mental characteristics deemed to be desirable.'

According to a recent UNESCO report on the genome issue, 'whatever the theoretical and scientific benefits of such [information-based] categorizations, research scientists must always have regard to a more holistic appreciation of human beings, considered both as individuals with an inherent dignity, and as communities living in a given environment and culture.'⋆

For many people, this may seem self-evident. An overwhelming number still share the view of American Indians, and many other indigenous groups, that certain so-called 'informational objects' simply cannot be privately owned; they are common to all. When the Guayami tribe discovered that two Americans had filed a patent application in which they claimed to have 'invented' the tribe's cells, its leader flew to Geneva to protest the claim. He said, 'I never imagined people would patent plants and animals. It's fundamentally immoral, contrary to the Guayami view of nature, and our place in it. To patent human material . . . to take human DNA and patent its products . . . that violates the integrity of life itself, and our deepest sense of morality.'[13] (The patent claim was withdrawn following opposition by, besides the Guayami General Congress, the World Council of Indigenous Peoples, the Rural Advancement Foundation International, and the World Council of Churches.)

⋆See UNESCO International Bioethics Committee, *Report of Subcommittee on Bioethics and Population Genetics*, final version, 15 November 1995: for many indigenous peoples, the dignity of their ancestors is 'in our blood, our hair, our mucus, our genes,' accordingly some research projects are seen as an unwelcome interference 'in a highly sacred domain of indigenous history, survival and commitment to future generations … The gene and genome are not the property of individuals but rather are part of the heritage of families, communities, tribes and entire indigenous nations.' A copy of the report can be viewed on the Web at http://www. biol.tsukuba.ac.jp/~macer/PG.html.

However, according to the late Ron Brown, the US commerce secretary, whose department subsequently filed patent applications on the cell lines of indigenous peoples from the Solomon Islands and Papua New Guinea, 'under our laws, as well as those of many countries, *subject matter* relating to human cells is patentable and *there is no provision for considerations relating to the source of the cells* that may be the subject of a patent application.'[14] Indeed, the uncomfortable truth is that the reclassification of the world into data renders *everything* potentially subject to new forms of information management. The process is sweeping the planet, and it is hardly confined to Indonesian villagers or Amazonian Indians. TRW, an American firm that ranks among the powerful consortia racing to develop satellite-based mobile-phone services, recently went so far as to seek a patent for *orbits* around the earth.

The issues raised by the Human Genome Project are also mirrored in the world of bio-engineering. In 1980 the US granted a patent for a microbe capable of digesting oil. But is it ethically acceptable for a profit-seeking enterprise to patent a naturally occurring plant extract that cures cold sores and warts? Or indeed any element of nature that has been modified by a reshuffling of DNA which has freely evolved over the millennia in a shared space? Where such resources have hitherto been regarded to be the common heritage of mankind, the very essence of nature is itself being reconceived in informational terms; at the same time, regulatory policies are rendering this new territory open to legal exploitation.★

★According to the UNESCO report (see previous footnote), Europe has so far resisted this trend. It refers to 'the draft European Convention of Bioethics, article 11, which states, "The human body and its parts shall not, as such, give rise to financial gain." However, in note 90, it is stated that this does not apply to discarded tissues, such as hair and nails, "the sale of which is not an affront to human dignity." This is important to note because DNA can be obtained from discarded tissues.'

Outriders into this new marketing frontier – mainly American pharmaceutical and biotech industries, together with their political protégés who granted that first 1980 patent on life – argue that patent protection is required to stimulate costly research and development. There is little evidence to support this claim. As a matter of fact, such patents seek to *reduce* the free flow of informational exchange, not least by binding scientists to agreements to maintain commercial confidentiality, and they *restrict* the fruits of that research to those with the plushest accounts. This process is being allowed to proceed despite the fact that the pre-informational system of public ownership of such 'data' has already yielded wide and demonstrable benefits for years.

Dr Jonas Edward Salk invented the polio vaccine and was rightly honored for the work that he did to relieve the suffering of humankind. Following his breakthrough, he was asked in a television interview who would actually *own* the new drug. Salk immediately replied, 'Well, the people, I should say. There *is* no patent. Could you patent the sun?'[15]

According to many digerati, the answer in principle (if not in practice) is a definite *yes*. Indeed, says Intel's Andy Grove, '*the economics of our industry only work if we have large numbers of users demanding our services* ... we need to be relentless in our efforts to increase the number of users and different uses of our technology.'[16] As Bill Gates cheerfully admits, the economic success of a cybernetic economy depends crucially upon 'total participation' in a digitized and fully networked landscape whose 'pervasiveness is part of the design.'[17] Those who choose to disconnect will automatically be marginalized.*

Many years ago, among the seafarers in the Lampung district on the Indonesian island of Sumatra, textiles played a central role in ceremonial life. Among the most beautiful of these were the so-called 'ship cloths' – gorgeous handcrafted tapestries of homespun cotton that were woven in rich reds, golden yellows, and blacks. They depicted an 'upper world' inhabited by birds and spirits. Floating on the middle sea of life they showed an ark-like vessel, loaded with passengers and beasts. Beneath were portrayed the creatures of the deep. Taken together, these suggested a kind of universal symmetry: a sense of balance between man and nature and of continuity amid oceanic change. These cloths were brought into the world exclusively by women in what can only be described as a sacramental rite. The women would gather on rattan mats laid in a jungle clearing, or on the sand by the palm-fringed sea, and would weave for timeless weeks before a single cloth emerged. These textiles were celebrated by Sumatran society, for they gave symbolic shape to the very warp and weft of its existence; they conjured the whole galaxy of rituals, customs, and traditional beliefs that defined the island's way. If there was a wedding or an enthronement, or transition ceremonies surrounding birth or death, these tapestries were unfolded and hung as backdrops, forming a sacred space in which the sacraments could proceed.

*If a technology becomes like an umbilical cord – when you cannot unplug or opt out without threatening your survival (economically or otherwise) – then you are truly a captive of that system. Ironically, cybernetic systems are often justified on the grounds that they introduce a greater 'freedom of choice.' But these are *conditioned* choices. When you call a number and get trapped in voice jail, you are offered a 'choice' between buttons to push, but rarely the opportunity to speak with another human being. In other words, the most vitally important option is denied.

143

This might help explain why ship cloths were deliberately withheld from trade, even if textiles were a leading means of exchange. In this archipelago, which straddled the main sea link between Asia and all other parts of the world, outlanders from China, India, Arabia, Europe, and Africa had met for centuries to trade resins, spices, rare dyes, and colorful bird plumes, as well as metal ingots of gold and silver. Cultural dialog was the stuff of Sumatran life. Invited into a distant Muslim temple, the clansmen knew to remove their shoes. Receiving guests from abroad, they entertained them in their own inimitable way. These dialogs required some sacrifice and accommodation by all concerned: they required an ability to *listen*. This is what made them enriching in a spiritual as well as a financial way. Trade may have been a golden strand, but it was nevertheless subordinate in the much richer tapestry of life.

Conventional theory states that the more involved a people become in an international dialog, the better off they should be.[18] This is an oversimplistic (if still dominant) view. As with individuals, so with cultures: the truth is that, in the wrong context, certain experiences can be too much to absorb. Instead of adapting and assimilating, one is simply overwhelmed. Instead of being able to redefine oneself amid the flux of change, one's situation is transformed by external events. In all cross-cultural 'dialogs,' everything depends on how the conversation is framed. Some Sumatrans did indeed flourish when the Europeans arrived – those who were quick to adopt the new relativities – just as some people in the developed world now thrive materially on 'borderless trade.' They were absorbed into the new elites. However, for the vast majority in Lampung, the encounter with the West carried a high price that was reflected, at the close of the last century, by an abrupt and mysterious end to the weaving of ship cloths. A profound malaise suddenly seemed to grip island culture, and it was directly linked to subordination under colonial rule.

The key word is *subordination* – an absence of real dialog between Sumatra and its new overlords from the West. The respectful removing of shoes in the temple, the weaving of cloths, and all the sacred rituals that bridged the distance between tribes were suddenly dismissed as quaint and time-consuming curiosities. The

Europeans were obsessed by a far more impatient mercantile imperative. To this end, they demanded a relationship in which the masters could broadcast but only rarely receive. Thus, within the strictly economic topography imposed by European rule, there was no space for Sumatra's *life code*. Its people were disconnected from the landscape of their dreams.

Conquest and colonialism, projected by the brute force of superior arms, were reinforced by the imposition of a new matrix of values, of a mental landscape in which the significations of the ship cloth could only unwind. For one thing, it simply made no sense to devote months to the weaving of ceremonial fabric: it was an unaffordable luxury. Time was a commodity now, and 'productive' work was required from the islander who hoped to survive. As the traditional leaderships were either co-opted or destroyed, ancient rituals were allowed to falter and lapse: they were transformed from lived ceremonies into empty spectacles exhibited for the pleasure of those who now ruled. In effect, a rich culture, possessed of the inner strength to face the unknown and to engage it in open exchange, was overwhelmed as a result of its world being *redefined*, quickly and unexpectedly, in ways it was unable to resist.[19] Another page of life-giving diversity and inherited wisdom had been effectively torn from the great volume of life.

Through the sweep of human history, technically advanced powers have always overwhelmed the weak and imposed their own values and world-views on the conquered tribes. Empires grow by redefining the existing boundaries – even those of culture and nature itself. When England's king Charles II ordered a survey of his American colonies in the seventeenth century, the resulting charts depicted a reality of forts, plantations, and parish boundaries, not sacred waterholes and native Indian settlements. The new maps reflected *imperial* motives and views. While the majority of American Indians felt humility toward nature – it was something awesome, to be revered and propitiated, and which no person could ever have the arrogance to 'possess' – those escaping Europe were driven by contrasting priorities of domination, exploitation, and control. They came; they mobilized superior technologies to conquer; then they proceeded to measure, rename, and carve up their new territories into parcels of privately owned land. In doing so, they claimed to be following biblical injunctions to tame heathen peoples and dominate the chaos of nature, little knowing what social and ecological chaos they would eventually unleash. Likewise, when the great powers colonized Africa, they divided the continent according to the availability of natural resources, and without regard for its carefully delineated tribal boundaries and animal grazing patterns. They, too, were quick to annex the old social space, to create a new cultural landscape, and to indoctrinate the divided populations in a fresh language (which embodied explicit value judgments in matters of both emphasis and exclusion) and in a new sense of affiliation and loyalty. The overlords were of course obliged to eliminate lifestyles that obstructed their ambition, but they usually justified these necessary expedients on the grounds of some higher destiny.*

*Empires rise and fall like the tides. History, Arnold Toynbee believed, is made not by nations but by civilizations. Islam, Christendom, Buddhism and Confucianism have outlasted the various empires in which they were contained. Each has been extensively hybridized, to be sure, but a basic spiritual geography remains. Lacking this underlying ethical cohesion, the cybernetic domains of the new 'overlay culture' have more in common with a transient empire than with a true civilization.

An identical procedure has been quietly instigated today. Like an imperial map, a networked system is being superimposed on the fabric of life. Binary code is being insinuated as the ultimate arbiter of reality, the metalanguage of all languages, just as the substrate of a silicon chip is the essential route which almost all descriptions of that reality now cross. Visual images, sound, olfactory and tactile stimulation: all of these are now – or soon will be – conveyed by an underlying use of binary machines. Ultimately, we will be offered total sensory immersion in electronic worlds. But already the focal point for the great majority of human activities – activities that once occurred against a free and shared physical backdrop – has been shifted onto virtual terrain: a landscape that is predominantly owned and operated by private enterprise. We are witnessing the ascent of what one AT&T executive calls a 'global overlay culture,' an overprivileged and wired elite composed of wealthy individuals, scientists, government decisionmakers, corporate leaders, and so-called 'symbolic analysts' who are most empowered by the tools and values of a networked world. This 'cosmopolitan, largely English-language' culture is 'driven by the desire for *efficiency, predictability, and uniformity*.'*

It is committed to a timeworn instinct to expand through conquest, not persuasion; to maintaining a momentum that is dangerously disengaged from the natural rhythms, disciplines, and limitations of earthly life, merely because this momentum is immediately profitable to those who happen to control the new tools. Streaking headlong toward a cybernetic dreamscape and disavowing any notion of physical boundaries, ignoring the infinitude of possibility that can exist within the limited intersections of time and space – indeed striving above all to *escape* those very limits – they are guiding us toward a cloud-capped Babel of mutually exclusive experiences, an abstract terrain of boundless desire that 'no longer matches the landscape of fact.'[20]

*Frederick S. Tipson; see the footnote on page 138. He writes that the power of this overlay culture 'has been greatly facilitated by the increased exposure of poorer and more closed societies to the images in richer countries of prosperity and prowess ... [through] intensified communications links.' He adds that one of its essential characteristics is *economic homogenization*: ' ... Reuters, Dow Jones, *The Economist, Financial Times,* CNN, Disney, Microsoft, and Sony typify the brand "icons" of a common elite culture driven by the desire for efficiency, predictability, and uniformity.'

The twining of code and cybernetic systems is giving rise not only to a new social space but to a new *perceptual* terrain – one whose descriptions bear little necessary resemblance to those of the natural world on which we depend. Indeed, as we will see in the next chapter, the most far-reaching of all the changes wrought by information technology revolve around a subtle reconfiguration of the context within which we relate to each other, the means by which we view the world, and the ways we describe what we see.

7
Culture versus System: Language

Values are the basis of a people's identity – their sense of peculiarity as members of the human race. All this is carried by language ... the collective memory bank of a people's experience in history ... language carries culture and culture carries ... the entire body of values by which we come to perceive ourselves and our place in the world.

NGUGI WA THIONG'O[1]

The Aborigines of Australia lived by a creation myth in which each feature of the surrounding country was interlinked. They told of

> ... legendary totemic beings who had wandered over the continent in the Dreamtime, singing out the name of everything that crossed their path – birds, animals, plants, rocks, waterholes – and so singing the world into existence ... no one in Australia was landless, since everyone inherited, as his or her private property, a stretch of the Ancestor's song and the stretch of country over which the song passed. A man's verses were his title deeds to territory. He could lend them to others. He could borrow other verses in return. The one thing he couldn't do was sell or get rid of them ... the structures of kinship reach out to all living men, to all his fellow creatures, and to the rivers, the rocks and the trees.[2]

Now and again, when the elders decided that it was time to repeat the song from start to finish, each tribesman would follow the

Ancestors' footsteps, sing his stretch of country in turn, and thus help to keep it alive.

Far away, in North America, the Pueblo Indians helped 'Father Sun' through the sky, convinced that, if they once lapsed in their modest rituals, the world itself would grow dark. These origin myths, and the languages that contained them, acted as vessels for each culture's inherited values and shared beliefs. When a northeastern stretch of Australia was renamed *Queens*land in the course of white settlement, the cultural implications were far-reaching indeed.* The Aboriginal universe was quite literally being remade.

Language is the subtle tool we use to colonize reality. To fire off a word or phrase is to impose a signification and emphasis all one's own. Leonardo Sciascia, a courageous Sicilian author, wrote in his *Mystery of Majorana* that 'names are more than a definition of things, they are the thing itself.'[3] As a scientist, the late Heinz Pagels thought it crucial to recognize that language is a *version* of reality: a simulation. He wrote that by 'creating a substitute we have but spun another thread in the web of our grand illusion.'[4] There is something unique about the linguistic illusion, however. Having evolved over time, and having been used by so many radically different personalities, it becomes a vessel for the values on which a culture can agree. It tends to favor that which can endure. Like a great work of literature, whose nuances may be interpreted by different people in as many diverse ways, its strength lies precisely in the fact that it nevertheless *draws people*

*King Charles V of France, in a widely quoted and politically incorrect quip, pronounced that God was best addressed in Spanish, lovers wooed in Italian, and horses commanded in German. Today, in France, instead of merely dialing someone's phone number, you *compose* it. Your hostess might well express *desolation* should you decline her invitation to dine. Small differences, but telling in their own way. Even the syntax (or basic structure) of a language can disguise unique priorities, as the Argentine writer Borges illustrated in a tongue-in-cheek tale about a fictional planet he called Tlön. Its metaphysically minded inhabitants have stripped their vocabulary of all nouns. Instead, those of the northern hemisphere use *verbs* exclusively. Among southerners, a string of *adjectives* is preferred. Thus, in the north of Tlön, 'the moon rose over the sea' would be expressed as 'upward, behind the onstreaming, it mooned.' A southerner might express this as 'airy-clear over dark-round.' See Jorge Luis Borges, 'Tlön, Uqbar, Orbis Tertius,' from *Ficciones* (London: Everyman's Library, 1993), pp. 10–11.

together, year after year, in their shared consent to its resonance and metaphorical rule.

A sensitivity to this central aspect of language imposes a certain modesty on anyone's pursuit of ultimate truth. Contemporary scientists appreciate how the very terms in which an experiment is framed will powerfully influence the actual result. But in seventeenth-century Europe a different mindscape prevailed: a rationalism that was as confident in its dominance as the spread of its written texts was swift.* This was the start of a secular and pragmatic phase, anxious to cast aside old rituals and unproven beliefs, and to dominate the world through exercise of the intellect. Like many of his kind, Gottfried von Leibniz, the German philosopher and mathematician, was convinced the world was ultimately knowable, that its fixed mechanisms were universal, and that it could all be described in a linear, systematic, and impartial way. He resolved to devise the perfect logical medium: a linguistic vessel that could contain everything *except* the relativities implicit in human speech.

*Marshall McLuhan argues that, as tribal storytelling gave way to the spread of typographic means of information exchange, the balance of the human senses was also profoundly changed in favor of visual, fragmented, *private* points of view. See his *The Gutenberg Galaxy: the Making of Typographic Man* (Toronto: The University of Toronto Press, 1962).

What Leibniz produced was the 'electric language' that we now call binary code. It is breathtakingly simple, even elegant in its mathematical purity. Using different permutations of zero and one, it can effectively describe, arrange, and calculate practically any facet of creation. Indeed, McLuhan wrote how 'Leibniz, that mathematical spirit, saw in the mystic elegance of the binary system that counts only in the zero and the one, the very *image* of creation.'[5]

As one communications specialist explains, 'Cybernetic theory and computer technology require rigorous but straightforward languages to permit translation into nonambiguous, special symbols that can be stored and utilized for statistical manipulations. The closed systems of formal logic proved ideal for this need.'[6]

Of course the essential meaning of any narrative – or linguistic system – is contained both in what it hides and in what it displays.

Jaron Lanier, the dreadlock-topped musician and virtual-reality pioneer, has observed that 'information is an *alienated* experience.'[7] It is about the representation of 'objects' that have been removed, or alienated, from their natural source – something like jewels plundered from the forehead of Buddha in a Tibetan shrine. Today, nevertheless, binary code is becoming the metalanguage of humanity, the symbolic system in which all of reality is framed. 'The cool universe of digitality has absorbed the world of metaphor,' writes the French thinker Jean Baudrillard.[8] And it is even more deceptive than most spoken human languages; its subjectivity is harder to recognize. Although it presents a convincing *illusion* of neutrality, it too embodies a definite set of priorities. What are these values? How will they influence the shape of the electronic landscape, condition the nature of its contents, and alter the overall mental terrain? Can an invisible and unspoken code actually redefine society and alienate it from the sustaining energies of life?

A strange dream floats through my mind. It has that irrational clarity of a painting by the surrealist Salvador Dali. In the dream, I see an immense, illuminated screen. Within its frame, I can see churning waves of a windblown sea. But, in the dream, I reach down behind the screen with my mind's eye into the depths of the machine itself and see a torrent of minute bits and bytes. Suddenly I realize that these are the binary units which give the screened image its shape. I have been confusing a highly filtered *rendition* for something real – an actual sea. I have been fooled by the interface – the screen. Yet, more and more, people describe and envision reality with the aid of an invisible language. It grows all too easy to forget the distinction between the sensual ocean that surges in the physical world and its coded representation produced by code behind a screen.

Living human culture – indeed all of nature – is far more complex than any programed routine can ever be. A real ocean is moved by a staggering interplay of energies, a grand dance of chance, cause, and effect. Likewise, within the overall unity of the human family, you will find distinct tribal climates, invisible

151

spiritual tides, and great surges of physical migration. It is these – and not simply programs or official policies – that have most profoundly shaped social experience to date. All are interlaced in complex and mysterious ways. Say a dolphin leaps toward the sun and then splashes back through a glittering wave. As a result of this microevent, the entire ocean is infinitesimally changed: its energies are subtly rearranged. If you build dikes along one stretch of coast, there may well be flooding elsewhere. Indeed, the most essential aspect of creation is that no single part of it can be isolated from the rest.* Modern scientists would put it like this: the universe is continuous, not 'granular'; there can be an infinitude of possibility even within the most limited space.

*Tolstoy wrote that ' ... for the human mind the absolute continuity of motion is inconceivable. The laws of motion of any kind only become comprehensible to man when he examines *units* of this motion, arbitrarily selected.' He added that 'at the same time it is from this arbitrary division of continuous motion into discontinuous units that a greater number of human errors proceeds.' (Leo Tolstoy, *War and Peace* (1863–9), trans. Constance Garnett (London: Heinemann (1904), 1976), p. 887. (My italics.))

The notion that everyone and everything are linked together in an essentially ethical web is shared by most of the world's great spiritual movements. An American Indian chieftain remarked early in the 1900s that 'a man who sat on the ground before his [dwelling] meditating on life and its meaning, accepting the kinship of all creatures, and acknowledging unity with the universe of things ... [would be] infusing into his being the true essence of civilization.'[9] In our time, Václav Havel wonders at the mystery whereby 'each spiritual act is an integral part of the order of the spirit, [and] the order of the spirit is present in each act just as the entire river is present in an eddy.'[10] According to Edward Said, a distinguished contemporary Arabist, our very survival in a complex and interdependent world is now utterly dependent on 'the *connections* between things.'[11] Of course, what counts is the living *quality*, not simply the coded *quantity*, that underlines these connections; herein lies the true distinction between communication and mere message exchange.

The virtual re-representation of reality is based on a profoundly contrary world-view. It proceeds within finite enclosures. It is defined at its most fundamental level by reducing the immensity of lived experience to a syntax of division and control. In order to work at all, the simulated world requires a fragmentation of natural continuity, its translation into granular units of binary code, and only then its reconstruction as a *computable* dream. (Such division and conquest is, of course, the time-honored *modus* employed by those seeking material or political gain.) The sound waves of a Mozart quintet are profitably transformed into assemblies of digital pulse. The Human Genome Project will capture an exploitable blueprint of human life in the encoded patterns of our DNA. The very oceans and skies can be re-presented and modeled in code. All of these bits swirl about behind the great cybernetic screen. A scientist interested in, say, fluid mechanics can take his virtual Pacific and infinitely manipulate, or 'reconfigure,' it at will. He can double or halve its size. He can drain it without its running dry. Creation is redefined as a coded collage that can be made to behave in ways that Nature, in her resonant complexity, would never permit. (Thus, the 'flexibility' and freedom conferred by the cybernetic system come at the expense of its accountability to physical (or analog) world rules.) Ironically, the ultimate aim of a cybernetic system is to successfully mimic nature: that is, for the simulation to become so convincing and so ubiquitous that it can be used to model, alter, or even *supplant* the real thing. Binary symbol manipulation is already being used to alter the very stuff of physical life, as the reader has seen.

Too often, however, the silent play of digits, tumbling through sequence and circuit, comprehends only the objective surface of things – those elements that can be unequivocally interpreted by the electronic systems in which they are contained. It cannot encompass the elusive meanings that beat at the heart of our ways; nor can it admit such troublesomely imprecise notions as *account-ability* or lust for power. It encourages us to objectify nature in terms of *signals* and to dismiss the animating spirit that pervades it as unintelligible *noise*. In fact, binary language was deliberately designed to muffle a great many historical, metaphorical, and spiritual resonances and to scatter any continuity of values that might lie at the core of our collective Being. Its hidden bias increasingly infects the way we speak and think. It is casting us into a spiritless desert, a place of false neutrality, an indeterminate territory in which we can take false flight from the necessarily subjective responsibilities of common choice in a grounded time and place. The world is redefined as a series of mathematical equations in which, as Hannah Arendt writes, 'all real relationships are dissolved into logical relations between man-made symbols.' This mathematically preconceived world is nevertheless another dream: a place that 'has the character of reality only as long as the dream lasts.' Indeed, what makes this landscape particularly remarkable is that, here, within its deceptive embrace, 'we deal only with the patterns of our own mind.'[12]*

*This is the very same place of desiccation where the poet T. S. Eliot saw 'Shape without form, shade without colour;' one where the human being is reduced to a 'Headpiece filled with straw.' See 'The Hollow Men' (1925) in T. S. Eliot, *Selected Poems* (London: Faber and Faber, 1954).

Communication, in its transcendent sense, implies the existence of a unity: the possibility of finding commonalities of outlook and purpose amidst diversity, of maintaining continuities amid oceanic change. The cybernetic system, in contrast, implies a codification of polarity, of discontinuity, and of *difference*. It begins by defining the terms with which we describe ourselves, by conditioning our expectations of each other, and ultimately by framing the context in which we can act on what we believe. It works on several linguistic and syntactic levels – some fairly obvious, others exquisitely elusive and subtle.

In the days when the majority of the world's population was engaged in farming, there was a certain universal resonance associated with 'laying a furrow through stony ground.' Although the analogy itself was trivial, it nevertheless reflected something important about the way people understood the challenges of their day: namely, that tilling the land and continued existence were one and the same. The industrial age brought new and more mechanical analogies: we were moved into a steel mill and invited instead to 'establish links' with our kind. Many of these analogies are still in everyday use. But more and more it is the paradigm of cybernetic systems that conditions our speech and patterns contemporary outlooks. In politics, people are invited to give their *input*, not their active participation, in what has now been redefined as an abstract *process*. Instead of being inter*related* (as in sentient creatures) people frequently speak of being inter*connected* (as in digital machines). The computer executive perceives customers as *users*. And it would seem that some users, exhausted by endless *reconfigurations* and continuous attempts to *optimize*, simply require more *downtime* than the rest. Still others wonder whether they are suited (or rather *designed*) for the *modes* of life on-line. But *Wired* magazine's 'Jargon Watch' column has already pronounced an ominous decree: should you be unwise enough to disconnect, you will automatically join the legions of *PONA* (an appalling acronym for Persons of No Account). Feel nauseous? *Wired* will permit you a brief 'bio break' in 'meatspace' (as if this could provide relief from a world in which human beings are assuming a dual role as both information *and* processors, and where the syntax of atomization for the sake of 'efficiency' prevails).

These hidden values are subtly advancing into public discourse. Michael Prowse, an outspoken commentator who writes in the *Financial Times*, recently offered the following modest proposal with regard to universities. Henceforth, they should be reorganized along the same production lines that have revolutionized Japanese factory life. Inevitably, he says, new modes of 'just-in-time electronic education, delivered to your living room by commercial companies, will undermine the most hallowed names in higher education.' We should welcome such a modular approach. After all, 'the plain truth is that [universities] are selling a *product* that is ridiculously expensive and ill-suited to the needs of a rapidly changing economy.' It would be a lot smarter to focus on developing the student's 'cognitive abilities' with respect to specific *entrepreneurial needs*.[13]

Prowse evaluates the university system on the narrow grounds of its functional and economic expediency, ignoring its place in a pluralistic society. The basic performance criteria applied to this system, its teachers, and its students are identical to those used to measure cybernetic machines – which is to say cheapness, quickness, and efficiency in the tasks at hand. What is demanded of the student is practical competence at clearly measurable and above all commercially useful tasks. Reinforcing what we already know or desire – the status quo – is given greater value than preparing ourselves for interactions with the unknown. And acquisition of *useful* data is given precedence over applying it with integrity or judgment. If the study of the humanities can enrich our collective experience, and help mold citizens capable of relating their cultural inheritance to contemporary problems of common concern, this is lost in the riptides of specialization and logic.* It seems somehow ironic and perverse in this age when people are expected to live by their wits in a maelstrom of change that they should be denied the best educational preparation to do so. Instead, we entertain notions of 'education' in which learning is broken off from the holistic human context which makes it meaningful: a form of 'training,' or indoctrination, in which our dialog with cultural history has been scrapped.

In the eighteenth century, such narrowly focused arguments as

*Dictatorial systems have always targeted universities – either by direct political oppression or by more subtle financial means. They do so for the simple reason that they have the most to fear from reason and critical analysis. People force-fed with any number of practical skills are not necessarily capable of imagining their implications on the overarching social terrain.

While privatized on-line education might be integrated into the overall education scheme at highly specialized levels in, say, engineering and applied sciences, total privatization of the university and indeed school system, as proposed by Michael Prowse, is an altogether different thing. It would dissolve the glue that binds these various practical skills to the human context in which they're used. The humanities are particularly valuable to society because they remind us of the essential distinction between knowing *how* and knowing *why*. The university invites us to join in the mainstream of human debate; it encourages its students to distinguish quality from quantity; to distinguish between satisfying one's personal inquisitiveness – pursuing individual desires – and asking questions of concern to all humankind.

Prowse doubts the 'relevance' of Plato's ideas to the challenges implicit in marketing mobile phones. Mobile phones will eventually go the way of steam engines, yet the debate about the true nature of 'reality' and 'illusion' will remain (if not intensify). Moreover, a familiarity with Plato's ideas, especially the arguments he marshaled to justify authoritarian rule, can forewarn any student destined to wield power in a democratic society, and create a sensitivity to the dangers and responsibilities associated with its use. Indeed, when a society is being exploded and atomized, it is doubly important that it nourish a wide interest in the first principles that lie at its core: only in this way can it vitally reconceive them to meet any challenges at hand. The eternal questions posed by the classics of literature, political philosophy, and fine arts, sometimes dubbed 'elitist,' are precisely those which help us define what we *are* as human beings. Without them, and without the ambition to excel as *citizens* as well as consumers, society would be lost. Higher education would have been reduced to a crude means of indoctrinating students in the codes and usages of raw conquest.

Prowse's could still elicit a devastatingly satirical response. Jonathan Swift, in a blistering tract entitled *A Modest Proposal*, suggested that the terrible poverty of Ireland could be alleviated by the simple expedient of selling its babies. Of the 120,000 infants 'produced' each year, four-fifths could be offered for sale 'to persons of quality' for use at their tables. At one year of age, he noted, they would be delicious 'whether stewed, roasted, baked, or boiled.' And the scheme offered several statistically demonstrable benefits, not just to the national economy but also to Irish parents and English consumers as well. Swift earnestly assured readers of his own 'disinterestedness' in the whole affair: after all, he had 'not one penny' to gain.

And yet the threat to universities is very real. Even as the university comes under politically motivated attack, a combination of financial pressure and technology is undermining many of its customary functions. As a result of the proliferation of knowledge, especially in scientific fields, scholars now cluster in specialized electronic communities as much as they do in the old, cross-disciplinary, islands of excellence they inhabited before. It becomes commensurately hard to integrate their work into a holistic view. Meanwhile, the cost of storing data in physical form, which is rising at the same exponential rate as information is being generated, means that no single university or library can be comprehensive in every field. It must reinvent itself as a *conduit*, a crossroads, a filter for digital information exchange. Here, as in many other parts of society, the timeless questions being raised afresh by the spread of digital tools include whom (or *what*) to trust in the role of inter-mediary, and what *values* should guide the way.

'In the past, people came to the information, which was stored at the university,' remarks one communications expert, Eli Noam. 'In the future, the information will come to the people, wherever they are. What then is the role of the university? Will it be more than a collection of remaining physical functions, such as the science laboratory and football team? Will the impact of electronics on the university be like that of printing on the medieval cathedral, ending its central role in information transfer?'[14]

The answer is in fact quite straightforward.

The university experience, like that of the church, the family, and the community, rests fundamentally on the communication of meanings situated in their human context, and not on the mere exchange of useful signals. 'Education is based on mentoring . . . role modeling . . . socialization . . . processes [in which] physical proximity plays an important role. Thus the strength of the future physical university lies less in pure information and more in college as a community . . . '

What's more, Noam adds, 'with the explosive growth in the production of knowledge, society requires *credible gatekeepers* of information, and has entrusted some of that function to universities and its resident experts [rather than] to information networks.'

You are a plainsman in southern Africa, or Australia, or somewhere in the vast desert spaces of the Taklimakan Shamo. Imagine that night has fallen, and that you are prodding a log on the fire. A sudden glint flickers at a high corner of your eye. Turning, you spy a pinpoint of light that is traversing the high vault of the sky. Once it passes on, the countless stars and planets glitter implacably away. With a shake of the head, you turn back to the flame, sensing your insignificance in the universal parade. Several thousand miles away, a different mood prevails. A team of elite technicians are huddled over a bank of computer monitors in their neon-lit control room. They are looking *down*, through the eyes of a remote-sensing satellite – the very same satellite that flashed so briefly at the corner of your eye. The technicians' screens are crowded with geologic data. The charts are tinged in poisonous pinks, oxidized greens, and noxious russet iodines. Using their orbiting sentinels, these technicians of the developed world are calibrating the density of mineral deposits, pinpointing the extent of underground water tables, and probing for new sources of fossil fuel. If these commodities exist in sufficient levels to merit exploitation, the owners of these tools will swing into action, approach the unknowing owners of the land, and the earthmoving behemoths will move in. They will bring roads, television. Other satellites will then beam down images promoting a miraculous new way of life. The lifestyle on the ground – among the receivers caught under the orbiting satellites' swath – will be irradiated by invisible forces beyond reach.

Whole categories of information that were traditionally considered to fall outside the ambit of the market system are now being drawn into its exacting embrace. Indeed, although the cybernetic revolution is often said to be built on enhancing communication, at root it is about redefining the world in informational terms, about increasing the volume of messages that can be *usefully* exchanged, and about exploiting the commercial opportunities associated with *managing* the resulting data flows. Music and video, translated into a series of numbers, can be arranged in lucrative patterns and distributed by those with the appropriate skills and tools. Factories can be run by numbers and networked across national bounds. By reducing all of the unmanageable multiplicities of existence to one universal system, it becomes possible to objectify the landscape of social and natural reality. When it is thus transformed into code, it is decoupled from the context that makes it meaningful and is thus traded as a mere commodity.

Because our perceptions and our rules have not yet caught up with this shift, many policymakers and executives have tried to mask its significance in meaningless euphemisms. (These serve to ensure 'the defense of the indefensible . . . to make lies sound truthful and murder respectable, and to give an appearance of solidity to pure wind.'[15]) Consider, for an instant, the specific trade-political language surrounding the so-called copyright industries: advertising, electronic entertainment, and computer software. These form one of the fastest-growing and most dynamic sectors of the world economy and are overwhelmingly dominated by the United States. This dominance is especially pronounced in audio-visual entertainment – a business with strong linguistic and cultural overtones. The large dream factories clustered on the American West Coast have targeted non-US markets as the main source of future expansion.*[16]

American news, specialized television programs, and video games claim an ever-swelling share of foreign spending as well.† America also enjoys a virtual monopoly on PC operating systems and boasts one of the greatest concentrations

*American-based producers are able to lucratively export their audio-visual offerings because they have already recovered production costs on the linguistically unified market at home. Filmmakers in other parts of the world – geographically, linguistically, and culturally fragmented – lack this advantage.

†According to the American Motion ↪

Picture Industry Association, the US copyright industries (which include advertising, films, TV programs, music, computer software, sound recordings, video and books) 'are one of America's largest and fastest growing economic assets,' generating $238 billion of US GDP. They 'contribute more to the US economy than any single manufacturing sector including aircraft and aircraft parts, primary metals, fabricated metals, electronic equipment, industrial machinery, food and kindred products and chemicals and allied products,' according to the International Intellectual Property Alliance's report entitled *Copyright Industries in the US Economy: 1977–1993* (see pp. iii–iv). They are 'a key component of the long-term prospects of the US economy.' America's audio-video exports were the the fastest-growing source of export revenue after motor vehicles in 1993, according to the US Commerce Department's *Industrial Outlook 1994*. In fact, the six major Hollywood-based film studios – Universal, Disney, Warner, Paramount, Fox, and Columbia/TriStar – together control 94 percent of the world market. The industry is rapidly consolidating. Time Warner and Fox (Murdoch/News Corporation) between them control half of US box-office receipts, states the London merchant bank S.G. Warburg (in its report *Multi-Media – Myth and Reality* (p. 53), dated October 1993, which also drew heavily on work by Paul Kegan Associates, Inc.). America's *foreign* revenues from filmed entertainment equal those generated within the United States. Warburg reports that in the 'longer term, the inexorable increase in demand for US product in India, Indonesia, China and East Europe will underpin [industry] growth.' Foreign sales of the US copyright industry totaled $45.8 billion in 1993.

of telecommunications power in the world.

Unsurprisingly, Washington has emerged as the leading exponent of a beguiling (and apparently indubitable) doctrine known as 'free information flow.' What is 'free information flow'? In a networked world economy, it is crucially allied with the 'free flow of goods and services' (because information has become a product in its own right).

But habits from an earlier time die hard: people commonly confuse the notion of free flow with freedom of speech. Pluralistically minded people rightly feel that speech should never be subject to political censorship in any form. But this is dangerously to misinterpret the word 'information' in its new cybernetic context, where virtually everything is perceived in binary terms. Indeed, the very notion of 'free information flow' is so broad as to be absolutely meaningless. The growing number of 'free flow' battles are in fact battles over information ownership, battles over data configuration, and disputes over its acceptable use. At core, they revolve around the question of who will 'capture' content and assume the role of intermediary – or *signal manager*.*

*One member of a cross-industry working team that is formulating common standards

In the copyright industries, these questions signify much more than

merely a scramble for fresh commercial markets. They imply a
sublimated attempt to redefine
the world's mental universe – to
devalue the art, history, geography,
and literature that form the basis of
the world's various cultures – and
to elevate the language and values of a colonizing cybernetic system
instead.* As Stewart Brand remarked in *The Media Lab*:

> Global consciousness is not everybody's idea of a good thing. Apart
> from the draining of national sovereignty inherent in the global cash
> register, there is the threat of the global jukebox and the global
> movie projector weakening cultural identities worldwide. Nothing,
> apart from physical home turf, is as ferociously defended as a group's
> unique sense of who it is and what constitutes right behavior. But
> the means of physical defense of territory are well known; the means
> of electronic communication defense have to be invented while the
> damage is being done, and all the skilled inventors work for the
> invaders.[17]

This awareness only stoked the
emotional flames that led to a short
circuit in the audio-visual negotiations between the EU and the
US in the winter of 1993. Wim
Wenders (an acclaimed German
director who made the film *Paris,
Texas*, among many others, and
who has worked extensively in the US) announced that 'there is a
war going on and the most powerful weapons are images and
sound.'[18] Bertrand Tavernier of France went further. He spoke of
cultural genocide: the Americans are 'treating us exactly as they did
their Indians,' he said.[19] Many independent US filmmakers are
inclined to agree – but doubt whether the problem is uniquely
American. They suggest the problem is actually a set of production
values that US-based studios happen best (but not exclusively) to
exemplify. (To speak of *American* film studios is deceptive, since

for America's 'information superhighway'
points out, 'If there's no way to get paid for
something and there's no commerce going
on, there's no system' (author conversations
with John Garrett at Corporation for National
Research Initiatives (CNRI), Reston, Virginia,
21 July 1994).

*In 1994, Jacques Delors, still president of
the European Commission, expressed his conviction that music and films 'are not just
commercial goods ... they are an expression
of the identity of each people, the vehicle for
their language, of their history, of their heritage
and their artistic patrimony' (James Pressley,
'EU stirs up controversy with paper on film
industry,' *Wall Street Journal*, 8 April 1994).

many have substantial non-US shareholders.) The main concern for an entirely market-driven media machine – US or otherwise – is to avoid taking undue risks or alienating the powers that be. The task is to satisfy financiers who are intent on maximizing the return on their capital employed.* Controlled by a new 'overlay culture' that speaks the universal language of the bottom line, these dream factories are characterized by the actor Gérard Depardieu as 'a war machine' – a state within a state.[20]

*US studios are often charged with producing work that is almost exclusively insular, with rote productions that tend to be short on depth, reflection, and human insight and which favor large, commercially dependable, and formulaic approaches to public distraction – work that is said to be accessibly episodic, undemanding, and quite thoroughly superficial. Does the recent popularity of cult and independent films suggest an emerging appetite for alternatives to this formulaic approach?

Hollywood has in fact captured over two-thirds of the French film audience and has built up an 80 percent market share in Europe overall. (In fact its expansion closely follows a pattern established by Microsoft in computer software: it conquers four-fifths of every market it targets and leaves survivors either to join in an orbit around the victorious sun or to risk being marginalized around perhaps more creative but less powerful Apple-like hubs.) Jacques Toubon, France's former culture minister, feels government has a duty to help its people enjoy a wider choice in its diet of films and TV programs, not just a selection from a narrow range that reflects 'the same model, the same state of mind, the same aesthetic.' Without proactive policy, the danger is that Europe (and other regions) will become tribalized, with its executives speaking English, ordinary people speaking 'the language of television – 400 words with every kind of sloppiness' – and intellectuals and public administrators speaking something resembling the original mother tongue. 'It would mean that our society ... would not only tear itself apart but the fragments would be incapable of understanding each other.'[21] As he put it to *Le Monde*, 'Anglo-Saxon countries ... are deploying considerable efforts ... to conquer new territory for their language.'[22]†

†On the opposite side of the Atlantic, many digerati speak of the 'unstoppable technical flood,' and seem to imply that, if something is

However financially self-interested the French containment policy may have been, a more profound issue inevitable, the pragmatic thinker is simply foolish to resist. Issues of responsibility, they seem to say, cannot be brought into play. (One can't help but wonder, in this world of 'objective' rule, whether there are still situations in which seemingly hopeless resistance is the only ethically acceptable course. The gradual rise of Hitler's brownshirts may have been 'inevitable' when you consider the confluence of historical and social circumstances that prevailed in Europe at that time. Still, why do we honor those who had the courage to ally themselves with the Resistance? Mere sentiment? It seems odd how the mainstream media tend to discount thoughtful challenges to similarly unpalatable (technical) trends today. Those promoting living culture are dismissed as 'elitist,' or, even worse, 'Luddite' reactionaries.) But there *is* nothing inevitable about the uptake of technology. For example, the present ascendancy of the private passenger car over public transport in many countries, far from being an inevitable outgrowth of internal combustion, actually followed from distinct economic and political *choices*.

lay at its core — one that was never fully credited in the discussions that took place. The dispute was not only between rival languages and commercial spheres of influence: it was about the sustainability of any living culture when confronted with the imperatives of a purely logical system — a dispute that smolders like a sleeping volcano throughout the world today. Back in the USA, meanwhile, in its article on the French film affair, *Wired* magazine produced a digitally composited image in which a ridiculous Mickey Mouse cap was placed on the late President François Mitterrand's bald pate.[23] It was a gesture of gratuitous mockery that spoke volumes about the narrowness of outlook with which many digerati interpreted the whole affair. Secure behind walls of crass and insulated intolerance, they propose with disingenuously simple logic that 'Films are a product just like any other. Like furniture. Or hamburgers. They are things we happen to make well.'[24] In 1993 a senior trade negotiator named Carla Hills suggested that the rational division of labor in a networked global economy would have France making cheese and Hollywood looking after enchantment. It was an amusing but quite serious quip. European quota proposals, which were aimed at limiting Hollywood inroads for an interim period so that European filmmakers could regroup, were even rejected as 'anti-democratic'; Europe was practicing 'government censorship.' Jack Valenti, that consummate power-broker who heads the American Motion Picture Industry Association and is sometimes known as 'Hollywood's consigliere,' shed crocodile tears over the fact that Europe

163

was blocking viewers from making 'their own decisions about what they want to see.' (More to the point, he warned, new technologies, including direct broadcasting by satellite and pay-per-view TV, will 'defy restrictive regulation. Barriers [and] protectionism are out of place in a world of *creative competition and expanding visual choice*.'*)[25] Mickey Kantor (then the official trade negotiator) added his concern that Europe's proposals 'would have enshrined the principle of limiting consumers' viewing rights.'[26]†

Note this subtle shift: Europe's *citizens* (who are presumably entitled to have social agendas of their own) are deftly redefined as viewing *consumers* (who desire nothing more than their *right* to an abundance of cheap and stimulating *product*). And only Hollywood – and its selfless political servants in Washington DC – is willing to attend to the world's vital desire for bread and endless circus.‡ One Parisian commentator, embroiled in a related dispute over Euro Disneyland, spoke for many when he wryly confessed, 'I've found it rather hard to follow the GATT [trade] dossier. All I know is that the naughty Americans, not content with having foisted Disneyland upon us, are now entrusting our future to a man named Mickey.'[27]

What is at stake is far more than imprecise notions of unrestricted freedom, or widened market share,

*In fact audio-visual 'choice' is being determined by an ever smaller number of communications conglomerates, according to Veronis, Suhler & Associates' *Communications Industry Report*, cited by Alice Rawsthorn in 'Search for a happy ending,' *Financial Times*, 30 December 1995. The concentration is being driven by 'fear that specialist film and music companies may find it increasingly difficult to sell their products if rivals are linked to the cable channels and television stations and online services owned by the same diversified communications groups.' They also fear difficulties in attracting recognizable brand-name (and thus bankable) talent plus essential managerial expertise.

†Rather than invalidate cultural concerns on trade-political grounds, *Wired* magazine takes a more subtle approach in its reportage on the US–French dispute. It says the real problem lies not with the Americans but with the French. Their entire, centralized social model is proving unworkable in its confrontation with the implacable 'decentralizing' forces of technology and the demands of a global market. Concern about commercial film is therefore merely a misplaced symptom of France's own deep malaise. The hard fact, *Wired* writes, is that 'Molière is no longer as famous as Shakespeare' (which of course suggests that literary quality is best judged on the basis of popularity alone). (John Andrews, 'Culture wars,' *Wired* (UK) 1.01, April 1995.)

‡Once the European threat was hacked aside, Americans turned their combined forces against Canadian 'cultural protectionism' in 1996.

or even dollars and cents. In the case of the news media, for instance, it concerns nothing less than how (and by whom) our collective dream is shaped. Art, like culture, cannot be expected to thrive when its value is measured in purely commercial terms. Consider the seemingly prosaic battle over the future of the UK's BBC TV, an outstanding network that consistently ranks among the most implicitly trusted of Britain's otherwise so unfortunately tarnished institutions. (A *Sunday Times* poll published on 3 December 1995 found that people regard the BBC as one of England's most 'civilizing' institutions.) The BBC system exists as an expression of distinct sociopolitical priorities. It emerged, phoenix-like, out of the rubble of the Second World War when citizens entered into a tacit social compact under which, for the price of a modest license fee, they were able to support the kind of quality and independence in programming that it would be impossible to obtain under a strictly commercial regime – in other words, the kind that helps, in small but untiring ways, to keep equally untiring authoritarian tendencies at bay. As a result, British subjects can now take pride in a network that 'shows a clean pair of heels to every other news and current affairs outfit in the world': one that advertises the very best values of the society that gave it shape.[28]

The BBC's intellectual integrity – its freedom from untoward political or commercial influence – is increasingly recognized as one of its greatest strategic assets. The BBC has earned a vast reserve of goodwill throughout the world – a reserve that remains invisible on its conventional balance sheet. At the same time, however, its fearless documentaries and hard-hitting news programs frequently contradict comfortable versions of reality put forward by many members of corporate and political elites, whether at home or abroad. In the UK, there are persistent calls that it be emasculated if not wholly privatized (which would result in advertisers and the business community having greater influence over what it has to say).* In the booming growth markets of the Far East, News Corporation chairman Rupert Murdoch summarily ejected BBC News from its place on the satellite transponder broadcasting Star TV

*Most recently, there have been calls for the BBC World Service to be reorganized on more commercial lines. P. D. James, the novelist and a former governor of the BBC, ⤶

spoke of the 'extraordinary arrogance' involved: 'I would like to say [to the new management] that they do not own the BBC. The World Service is not a private company. It belongs to this country and the people of this country and the people of the world' (In brief: 'Chairman defends standards at BBC,' *Financial Times*, 19 August 1996).

over the Asian landmass. As he did so, in 1994, he cited pressures from Beijing's octogenarian rulers: he wished to satisfy their delicate authoritarian sensibilities and was anxious to safeguard his strategic access to the vast market of poten-

tial Chinese consumers. The justification? Strictly business: ' . . . we said in order to get in there and get accepted, we'll cut the BBC.' Yet, only months before, the same Rupert Murdoch was entertaining a different audience at a speech in London with a gratifying (if Cinderella-like) tale of how advanced technology will prove 'an unambiguous threat to totalitarian regimes everywhere.'[29]

The uncomfortable fact is that information technology is often being used to project an altogether different agenda: the *realpolitik* of raw expansion. The filtering imposed by Mr Murdoch's News Corporation sends an unambiguous signal to those who believe in genuine pluralism and a true competition of ideas. For all the talk of an 'inextricable connection between free flow and democracy,' it suggests that, even in a digital world, it is trivially easy to both prosper and suppress. Indeed, an environment of instantaneous, transnational, and electronically mediated information distribution can also be one of more powerfully centralized news *control*.*

Or consider another instance, in which members of the European Parliament tried, on principle, to restrain the type and the quantity of advertisements aired during broadcasts aimed primarily at small children. The World Federation of Advertisers fired back with the argument that such ads 'provide the economic lifeblood to the [European] Community, generate jobs, [and] offer consumer choice as well as vital resources to fund both the

*Placing national public broadcasting systems on a sound economic footing (together with schools, libraries, and universities) and devising ways to make private database resources available for educational use count among the greatest social priorities in a time of transnational information flows. The 'free-market' approach to public good, often cited by those who oppose a strong publicly funded alternative to commercial media, will only result in further concentration and commercialization at the expense of the social system as a whole. Note that this approach has already produced such debacles as the American savings and loan scandal (which was eventually sorted out at great public expense).

traditional and new media.' The Council of Ministers in Brussels subsequently reversed the parliament's plans.[30] They, too, seem to have adopted a deceptively neutral language of expediency which can conveniently mask the moral content of the hidden choices. This is why contemporary advocates of 'free information flow' call to mind the immortal Milo Minderbender, that scoundrel in Joseph Heller's black farce *Catch-22*. Like Minderbender, they resolutely confuse the public interest with the considerably narrower priorities of their own private enterprise. Then, with further cheek, they proclaim they are selflessly running organizations in which 'everyone has a share.'

The democratic philosophers of the seventeenth and eighteenth centuries – often writing at great personal risk – firmly established the notion that sovereignty belongs in the hands of the citizen. One of the central tenets of their sense of democratic order was the idea that government should never be allowed to condition public perception by exercising a stranglehold over the exchange of ideas. Democratic thinkers of the eighteenth century like Edmund Burke, David Hume, and Thomas Jefferson understood that political liberty depended crucially on an informed citizenry (even if citizens were often, at that time, defined as white male property owners), and that this was best cultivated by guaranteeing a free press operating within a shared civic space.*

However, those philosophers worked in a very different time of pamphleteers: a time when ideas were freely distributed by handbills and word of mouth. Human networks were more organic and built to a different scale: they revolved around market halls, public squares, and pubs. But how do people arrive at a sense of rational self-interest in the very different networked world that prevails today –

*In his *Of the First Principles of Government*, published in 1758, David Hume marveled at 'the implicit submission with which men resign their own sentiments and passions to those of their rulers. When we inquire by what means this wonder is effected, we shall find that, as Force is always on the side of the governed, the governors have nothing to support them but opinion. It is, therefore, on opinion only that government is founded; and this maxim extends to the most despotic and most military governments as well as to the most free and most popular.'

one of sound bites, manufactured personalities, and mythic lifestyles; one in which the consumption of certain ideas and commodities has been invested with such imperial glamor as to make most other concerns seem irrelevant by comparison? Living in a networked landscape, the very lens filter through which we view most events has long been affixed to the problematic system itself. In many journalistic newsrooms, for example, the required skill sets have increasingly more to do with *packaging and marketing* information than with ensuring the integrity of the data conveyed. Hiding behind the cloak of a false 'objectivity,' and intent on exposing factual contradictions, the commercial news media too often fail in the more important task of providing critical analysis and seeking *continuities of truth*. Being captive to and answerable to the priorities of their advertisers, they claim to be serving the public interest by providing 'choice' when they are in fact primarily serving their own. The result, remarks Margaret Carlson of *Time* magazine, a popular panel-show performer on US television, is that 'The less you know about something, the better off you are.'[31]

At this phase in our evolution, government is no longer the main threat to democratic pluralism and a free flow of ideas. The developed world lives instead under a subtle tyranny of instant opinion polls, media tycoons, and subservient political hacks. The information space, of which the Internet remains a relatively small (and rapidly commercializing) galaxy, has been captured by business empires whose programs are too often antithetical to the interests of human society at large. What's more, citizens now face the added danger that it has become technically possible simultaneously to control the avenues of message distribution as well as the content of the messages themselves: in short, to powerfully condition the entire public information domain. The equivalent, in the days of pamphleteering, would have been a situation in which one self-interested minority in society simultaneously owned the pub, the public markets, and squares, plus the printing presses, and also employed most of the people who generated text.

Indeed, the media of the developed world are more commercially self-censoring than politically censored against. A great many (but not all) newsmakers have lost their analytical zeal and instead offer up sensationalized, commercialized, and/or distorted pictures of reality that further their own specific aims.[32] As Frank Rich, an editorialist with the *New York Times*, writes:

> TV journalism at its most insidious has less to do with content per se than with the market forces of the culture it inhabits ... the driving principles behind TV news magazines and contentious panel shows are the same that guide the rest of TV entertainment. Journalists ... become continuing characters, with exaggerated personality traits as rigid as the cartoon figures in *I Love Lucy* or *The Flintstones*. The shows' "plots," which emphasize confrontation and glib opinion-mongering, are similarly formulaic, whether the topic is O. J. or Bosnia. Journalism is not only short-changed but completely irrelevant.

The danger, he notes, is that 'a public estranged from the press is also disengaged from the institutions and newsmakers that journalists cover – and will probably look outside the system for both information *and* leadership ... [Today] the media are missing the story of their own role in compounding the alienation and anger that define the destabilized political culture of our time.'[33]

At the height of Athenian democracy, when a speaker rose to address the assembly, he was obliged to don a wreath of laurel leaves. In this modest way, he publicly acknowledged both the honor and the responsibility associated with citizenship. It is often said that new networking technologies could potentially be mobilized to revive such a civic consciousness. They could be configured to encourage 'spaces' of public assembly, deliberation, and debate, where citizens could relearn the arts of persuasion (which require that individual opinions be justified by an appeal to the common good). But, as long as the Internet remains a tiny side-show in relation to the information space at large, a more immediate and imperative priority – however ironic, controversial, and low-tech it may seem – is to place *public broadcasting* on a sounder financial footing and give it the independence it needs.

The truth is that, 'in the battle between the technologically possible and the economically attractive, economics always wins.'[34] (Actually, this is something of an oversimplification. The regulatory atmosphere and social climate are critical too.) The crucial confluence of commercial priorities and technical possibilities is reconfiguring the information space of modern society into a booming coliseum of colorful entertainments and anesthetizing distraction; it is creating a psychically *disconnected* commonwealth in which modest civic experiments are nearly always drowned. George Kennan, the former US diplomat who created the Cold War policy of Soviet containment, sees an 'abandonment by our government of much of the process of public communication, in education as in entertainment, to the good graces of advertisers, to people, that is, who have no public commitment, educational, intellectual, aesthetic or otherwise.'[35]

With the Cold War at an end, censorship by the commissar is being replaced by that by the accountant and 'free-market forces.' That, at least, was the view of Krzysztof Kieslowski, the late and highly celebrated Polish filmmaker who spoke with the unique experience of having worked both under the authoritarian oppressions of Communism (where people memorized banned poems and books to keep them alive) and in the markets of the West. Although Kieslowski personally prevailed against the toughest odds on both sides, especially with the successful production of his brilliant *Three Colors* film trilogy, one of his main preoccupations at the time of his untimely death in 1996 was the danger posed by 'so-called economic freedom.' It threatened the continued existence of independent thinkers and talented artists worldwide. Powerfully projected by informational tools, it could too easily give rise to a form of *market authoritarianism* in which free speech – and people themselves – are destroyed.[36]

All of world trade is increasingly guided by false 'freedom' and destructive cybernetic 'connections' of the kind noted above. The dehumanized and deceptively neutral language of 'free flow' addresses the superficial content of business environments in isolation from the *context* of life. It is reorganizing the world in ways that are profoundly undermining to the vitality and diversity of the world's cultural pool.

Consider another, seemingly unrelated, example: the successful attempt by American rice producers to break the Japanese market. During the dispute between Tokyo and Washington in the early 1990s, US government trade officials discussed the issue in legal niceties that boiled down to how many sacks of grain the Japanese would be obliged to let in. This choice of emphasis enabled the Americans to exclude the cultural significations at stake. Of course they knew that rice cultivation lies at the heart of many traditional Japanese ways. Food is considered to have a soul in the Shinto and Buddhist cosmology. Each year, Japan's emperor still ritually plants the first shoots of the harvest by hand, and the entire growth cycle is surrounded with elaborate ceremonies of prayer and thanksgiving. Rice cultivation is bound up with a lifestyle and an outlook: it has a resonance that cannot be captured by the price and quantity of output alone.

Now, traditionally, it was organized around small plots of two acres or less, which required continual tending and thus a close and extended engagement with the land. For an island nation chronically short of basic commodities, this guaranteed a certain natural resiliency, although it also carried a definite price. Small Japanese farms are radically 'inefficient' when compared with farms in the US, where tracts tend on average to be a thousand times their size. Because they lack the same short-term economies of scale, their harvest is commensurately less cheap. Of course, the low sticker price of import rice excludes important long-term costs associated with its production: namely, dependence on industrial-scale monocropping that not only reduces species diversity but encourages overcentralization, lowers resistance to crop diseases and pests, and forces an overreliance on high-tech chemical inputs.*

The quantitative price reflects nothing so much as the narrowness of the values used in producing it. Unfortunately, by the time people appreciate the extent of this fact, the Japanese farmers exposed to competition under 'liberalization' will have long since failed. The market for domestic rice will

*Just three crops (wheat, maize, and rice) provide two-thirds of the human population's dietary energy. Some 75 percent of all plant varieties have been extinguished during this century, due to industrial monocropping, and current gene banks are 'becoming gene morgues.' (See Geoff Tansey, 'Crop defenders gather in Leipzig,' *Financial Times*, 14 June 1996.) Meanwhile, according to the UN's Food and Agriculture Organization, world ⌐

food output will have to be increased by 60 percent over the next twenty-five years. The best way of achieving this goal, which even the World Bank has now recognized, is by moving away from the diseconomies of plantation-style growing and instead increasing the yields on small family-owned farms. This necessary shift in emphasis is politically and economically charged, needless to say. Yet, whether in plants, animals, or entire cultures, the destruction of species clearly produces its high short-term yields at a long-term cost. It creates a vulnerable harvest and draws down the carrying-capacity of the land. Thus, living generations inexcusably rob from those to follow.

weaken; plots will be ignored and fall into disrepair; land values will drop; the fabric of village life will start to unwind. Thus the availability of cheaper rice to Japanese consumers (a quantity that can be easily measured and computed) cannot begin to compensate for the dispersal of a life-giving cultural heritage – another of those elusive qualities which cannot be as easily subsumed into marketing terms.*

*The story would be incomplete without taking note of another more prosaic reality that underlay the above-mentioned dispute: namely, that Japan's protection of the rice market was an attempt to preserve the post-war financial relationship between the farm vote, which is disproportionately strong in parliament, and LDP politicians. It was this power-base, as much as respect for tradition, that supported the system that is now coming undone.

For all the loose talk of how the information 'revolution' will empower the common man, the greatest benefits of a globalized information economy are in fact flowing almost exclusively to those who are managing its myths, its technologies and its 'content.' Colonial 'charters' are being granted to the *signal managers* so that they can exploit the informational frontier in a (nominally) legal way. Thus the familiar distinction between First and Third Worlds, between North and South, is no longer merely geographical. It is also, increasingly, informational: it splits rich and poor countries alike. Those owning a PC can hardly compete on the same terms with the immense resources, and the infinitely more powerful dedicated processing systems, that are being mobilized by wired cybernetic enterprises. Of course, widened access to more information and an ability globally to coordinate action via the Internet are each helpful in their own way. However, the pace and stakes of the game have escalated dramatically, and it is doubtful whether the *relative* power balance has actually changed. There will always be a market to provide the wealthiest players with tools to stay ahead of (and therefore attempt to manage) unfolding events. Information is money, and the most relevant information becomes even more scarce and valuable in an information-saturated economy where response time is an important element in competitive success.

Such a market can be resisted only if its root values are re-conceived and the inescapable connection between our different cultures (and between ourselves and the landscapes of the earth) is fully embraced. A true communications revolution will require the cultivation of what American naturalist Barry Lopez calls 'a more equitable set of relationships with all we have subjugated.'[37] At heart, it lies in accepting the sacrifices inevitably asked of those wishing truly to connect.

Too many information-age warriors argue that, in a complex world, the only things we can agree upon are 'objective' facts. They show themselves to be trapped in the outdated frame of untempered seventeenth-century European rationalism. Evading the main thrust of contemporary scientific insight as well as the inherited spiritual wisdom of centuries, they take refuge in a false universalism, in a least common denominator, in tangible numbers that supposedly never lie. These are the values of binary code. And it is remarkable how often the language of small digits coincides with the imperatives of big money; how frequently the idioms of informational 'free flow' precisely coincide with those of unbridled and unacknowledged conquest.

The nuances of individual and cultural existence cannot, how-ever, be pushed aside indefinitely. A civilization cannot be built on a rabble of users and consumers alone. The longer human values are excluded from the stated aspirations of the informational globalists, the greater will be the world's suppressed pressures, and the more fierce and frequent the explosions of protectionism, xenophobia, and mass unrest. In a climate of unrestrained competi-tion for world markets, cybernetic business systems are lacking in the foresight, the corporate will, and the institutional capacity to act in sustainable ways. They are seemingly hard-wired to protect and enlarge their 'sovereign information space.' They are exhibit-ing the same behavior that characterized the undignified scramble by nation-states for control of colonial resources and physical terrain. The danger is that unless the new elites wake up to a shared long-term interest – which must be supported by informed electorates, a minimal level of social cohesion, and a greater sharing of social costs – they will destroy the very system upon which their prosperity depends.

Communication, whether between individuals, between cultures, or indeed between humankind and the earth, requires an inclination to *accommodate* rather than eliminate diversity – a common-sense restraint that respects different ways of seeing and being. It requires a new and more humane vocabulary which admits the obligations and duties implicit in the exercise of power, and a radically revised concept of the 'good life' in which nature is revered as a source not only of sustenance but of spiritual inspiration as well.

But the mythic landscape of cybernetic society differs profoundly from any other that has appeared on the human horizon to date. Its contours are not obliged to coincide, at fixed points in time and space, with those of the natural world: its borders are shaped by our dreams. Yet it would seem that information networking is being used to project the dreams of only a privileged few. Indeed, it is being used to erect powerful new *free states* – networked communication hubs that exist in parallel to existing political structures and regimes. Sometimes these free states vie with the old regimes; usually they comfortably coexist. But, more and more, national government – that sometimes lumbering creature that moves within the finite boundaries of geographical terrain, and works under *constitutional* restraints – is finding itself cast in the role of 'outdated dinosaur' in the adrenalin-charged domains of cybernetic power. The most electrifying concern is that the celebrated 'freedom' from physical geography will place the very institutions of representative democracy at risk.

8

Remote Control

One has to pay ... dearly for things. The dream of a freer, more meaningful
life is no longer just a matter of running away from Mommy ... but of a tough-
minded, everyday confrontation with the dark powers of the new age.

VÁCLAV HAVEL[1]

When young South African musicians sang of revolution in the late
1980s, they were singing about freedom from an apartheid regime
obsessed with building prisons instead of schools and hospitals, with
setting one black tribe against the next, and dividing Indians from
the coloreds from the whites. They were singing with the moral
force of a people unjustly oppressed by an illegitimate regime.
They were articulating an alternative vision of how *they* wished to
order their collective destiny.

The energies that swept South Africa onto its remarkable course
of political reform are very different from those driving the cybernetic 'revolution' today. When digerati rhapsodize about revolution, their song has an altogether different ring. They believe
that ubiquitous electronic networking will trigger a fundamental
reconfiguration in the balance of political power. The world
now operates with such speed and complexity, they say, that
formal public policy intervention is hopelessly late and almost
always ineffective. In a fully networked world, our collective
destiny is best guided by an invisible hand of economic liberalism.
Free trade, a free flow of information, and unfettered technical

development are democratic and 'liberating' forces too. Besides, any country that seriously hopes to survive into the twenty-first century as an economic power has simply got to set its individual and entrepreneurial energies free. This is the last remaining task of public policy.

The leading voices in this rather one-sided chorus of 'revolutionaries' are primarily white, almost exclusively male, and mostly North American. They include men like George Gilder, a technical 'visionary' associated with Newt Gingrich's ultra-libertarian Progress and Freedom Foundation, Kevin Kelly, the influential voice behind *Wired* magazine, and a host of New Age business and financial leaders. A good number of them – children of the 1960s acid culture now coming of political and entrepreneurial age – lived through the assassinations of President Kennedy and Martin Luther King. They also directly experienced the terrible escalation of the Vietnam War. Many, unsurprisingly, feel a deep-seated ambivalence toward governmental authority centered in Washington DC. But, whereas they set out wanting to start a whole new mode of living in America, many have since adopted more narrowly materialistic concerns. Their nonconformist energies are now expressed in a generalized sense that a rapid-fire, cybernetically wired world is no place for stodgy statesmen, inflexible judges, tiresome professors, and conciliatory diplomats. Events are unfolding with the lightning bursts of a video game. Only reflexive digital road warriors are fit and limber enough to hack through a chaotic and competitive fray. Only *they* will have the disinterested strength to prosecute the purifying struggles for commercial market dominance, to bring new cutting-edge products to life, and thus to guarantee the (economic) salvation of man by technical means. Some of the more restive outriders in this pack find it hard to conceal their outright contempt for almost any regulatory restraint. John Malone, a wealthy cable-TV magnate, heaps scorn on the very existence of elected lawmakers. 'I'd rather cut my leg off than go to Washington,' he rumbles at the interviewer from *Wired* magazine. 'I have no respect for politicians and I'm very poor at suppressing that.'[2]

It's nice to think that digital networking will initiate a process of fundamental reform in democracy. The status quo and its guardians can no longer claim basic legitimacy, after all. There is a widespread realization, as Gore Vidal writes, that 'the floor to this ramshackle civilization we have built cannot bear much longer our weight.'³ On every continent, news of financial malfeasance, corporate influence peddling, and skulduggery has become numbingly commonplace. Jean-Marie Guéhenno, formerly head of policy planning at the Foreign Ministry in Paris and France's ambassador to the European Union (EU), is convinced that the current level of corruption and wrongdoing that pervades the world's political class is a direct consequence of the information revolution, global networking, and the internationalization of markets without countervailing procedures to make the new powers accountable. He warns of an impending 'end to democracy,' unless we can 'rediscover that a human community is not only a political notion but a philosophical and religious one. Having lost the comfort of our geographical boundaries, we must in effect rediscover what creates the bonds between humans that constitute a community.'⁴

Yet inspired leaders young and old, confronting the tawdry spiral that public life has become, are losing courage and opting out. The halls of power are swarming with cynical ideologues and well-shod lobbyists. Speaking in dry whispers reserved exclusively for use among themselves, they exchange one gold-plated favor for the next. Publicly, meanwhile, they trade in adversarial polemics and sensationalism that feed a news mill at the cost of marginalizing reasoned debate. It seems as if the real decisions are taken in conclaves beyond public reach. If only citizens would cry out for a new script and players. Instead, they turn away from the whole public spectacle with weary contempt. This is disturbed only by their encounters with officialdom – which almost universally arouse the exasperation of a descent into hell. No wonder political commitment has dried up. It would seem as if the very concept of a shared public enterprise has gone with the wind.

The great irony about the digital 'revolution' is the way that these strong emotions – alienation, exasperation, a widespread desire for political 'change' – are being subtly hijacked and redirected to reinforce rather than displace the status quo. An emerging elite advocates a species of technically driven laissez-faire that promises 'empowerment' though it will actually further disfranchise the public at large. In earlier chapters we saw how the primary effect of cybernetic networking has been to shift the decisive context of commercial activity. The focal point of the business organization now resides more in software topology than in national geography. More and more, human livelihoods are derived not from nature but from the Net. But what are the implications of a networked digital economy that is increasingly disconnected from the 'analog' social and political institutions of physical space?

Democracy is founded on the basic proposition that people have the right to determine their own destiny. In order to do this, they have to regard themselves as more than just economic beings (or consumers). There has to be an active sense of citizenship, and a preparedness to confront or overthrow any regime that denies the public's ultimate primacy. Today, however, we have seen how life is influenced more by the coding, the configuration, and the operation of cybernetic systems than it is by traditional electoral means. The digital 'revolutionaries' ask that we embrace this as a positive fact. A wired economy, freed from unnecessary public policy restraint, will actually prove to be the best guarantor of the public good, proponents suggest.*

*In fact one of the towering absurdities of the digital 'revolution' is the way in which its proponents have so thoroughly ignored the observable if paradoxical fact that markets can work efficiently *only* when they operate under a framework of public oversight and occasional restraint.

Wired's editor recently produced a 500-page tome called *Out of Control*. Often fascinating, it tries to lend scientific credibility to this underlying anti-regulatory view. Mr Kelly thinks that we should pattern our lives on those of the lower insects. He writes that a complex system of any kind (whether it happens to be a bee colony, a financial market, or indeed a human society undergoing a process of cybernetically induced change) can be left to follow its own 'hive mind.' In fact, in a chapter entitled 'The Nine Laws of God,' he suggests that 'in a territory of rapid, massive and heterogeneous change, *only* a mob can steer.'[5] The entire regulatory edifice should be swept away. Two other digerati (also writing in *Wired*) predict a market-driven future of

'uninhibited, unstructured, undisciplined anarchy' where government is 'not only less helpful, but less relevant.'* This radical
deregulatory wing proposes not a *better* system of administration;
not a relayering of government responsibilities to reflect a more
complex world; not a reconfiguration of outdated representative
procedures in order to restore the balance of power in favor of citizens. It wants to eliminate government once and for all.

To advance this notion of 'free-market anarchy,' one has to be either disingenuous or naive. The very concept is not just a seductive fiction but an outright contradiction in terms. Free markets have never existed, and probably never will.† This is because an economy, at its most fundamental level, will flourish only in a climate of basic trust. This, in turn, requires a

*The last two quotes are extracted from an article by Don Peppers and Martha Rogers in *Wired* (UK) 1.01, April 1995. What an irony! The entrepreneurial digerati seem to be promoting notions that spring directly from Marxist thought – in which the state is seen as a necessary evil that will ultimately be destroyed. This belief contrasts with the democratic tradition of Locke, Hobbes, Montesquieu, and Rousseau – which is primarily concerned with the practical details of managing a state in the interests of society at large. During the Cold War, there were two officially sanctioned avenues to redemption. Under Marxism, it was through revolution. In the West, it was via high technology and entrepreneurial growth. Now we can supposedly pursue both revolution *and* entrepreneurial growth by technical means.

†For example, the prices of oil and its derivatives – commodities that stand at the heart of the modern lifestyle and which are produced by the most global of all industries – do not reflect anything like their true cost, being artificially depressed by both action (such as unneeded tax incentives that benefit producers) and inaction (such as failure to tax usage at sufficiently high rates) on the part of governments. This, in turn, reflects the political clout of petrochemical interests. Alternative energy sources, which are cheaper and cleaner to the environment, are routinely dismissed as 'uneconomical' when in fact they cannot compete on equal terms. Similarly, Japan and the United States, which between them control two-thirds of the world supply in computer chips, which lie at the heart of all computing products, have subjected their trade in these chips to two bilateral agreements. The first, signed in 1986, propped up prices at a time of oversupply. The second deal doubled the 'allowable' US market share in Japan to 20 percent. The result has been to force a long-term relationship between US chip suppliers and Japanese firms that design and produce electronic products. In mid-summer 1996 a third such pact was agreed. The basic principle of 'managed trade' underpins the world market in many other important commodities, including agriculture, textile goods and clothing, tropical products, leather and footwear, and many more. Too frequently these markets are 'managed' in favor of the well-to-do at the expense of the less fortunate.

comprehensible legal system, an agreement by citizens to limit their behavior in mutually beneficial ways, and a means of enforcement when legal codes and social conventions are transgressed. This is particularly true on the Internet, although it, too, is sometimes misconstrued as the penultimate expression of 'creative anarchy.' Even Howard Rheingold, a talented writer who has explored many of its fascinating byways in a book called *Virtual Communities*, describes it as an 'anarchic, unkillable, censorship-resistant, aggressively noncommercial, voraciously growing conversation among millions of people in dozens of countries.'[6] Internet aficionados, of whom I am one, point with hope to the Net's great potential as a creative force. But noncommercial? Anarchic? No way! For one thing, the launch of the Internet, and its remarkable rise into the high-flying reaches of business development, was initially triggered by *public* funding. Now it is managed by a group of telecom companies. Its continued operation depends not only upon their physical infrastructure but upon a series of distinct software protocols as well.* Without these,

*Internet communication depends on the TCP/IP protocol which connects many computers around the world. The network is woven together by the Internet Backbone. The physical infrastructure (wires, fiber optics, switching equipment) is owned by the telecommunications companies, which lease access to Internet service providers. Computers called 'servers', which are connected to the Internet, store the informational content. Computers that retrieve such information are called 'clients'. Most of the server computers run UNIX as the operating system, which

there *would* be no free-wheeling exchange. As with the Internet, so with the economy, and so with society at large. Each rests on rules, on infrastructure, on fundamental trust, and on workable means of intervention and enforcement when the systems fail. This quaint little fact is very often disregarded amid the storm of digital hype. In fact, however, life itself is built

allows multiple users to access a single server. The US government funded the original backbone via the NSFNet (National Science Foundation) from 1986 through April 1995. Merit Network, Inc., along with Advanced Network Services (ANS), IBM, MCI, and the State of Michigan, operated the NSFNet for the last seven of those years. Originally intended for non-commercial use, the network progressed in speed from 56 kb/s to T-1 (1.5 mb/s) to T-3 (45 mb/s). In 1993 the government halted its own funding and began a new network called vBNS, operating at (OC-3) 155 mb/s. The commercial portion of the Internet connects to the vBNS at various network access points (NAPs) and runs somewhat autonomously from it. As per April 1995, the Internet was commercially operated. The NSF funds the vBNS, which is operated by MCI. The routing arbiter is operated by on *cartels*. If thoughtfully designed and realistically adapted, they keep social and/or economic chaos at bay.

The word 'cartel' normally conjures an image of the proverbial smoke-filled room. The authors Daniel Burstein and David Kline use it to describe the US auto industry several decades ago:

Merit, and the NAPs are operated by PacBell (San Francisco), Ameritech (Chicago), Sprint (New York), and MFS (Washington DC). The next tiers within the Internet are the Internet service providers (ISPs) which are connected through the NAPs. ISPs have points of presence (POPs) around the US to provide local access to users of the Internet via either dial-up or leased-line accounts. Source and further info: http://www.nav.com/ ~lewallen/bkbone.htm

> Through a combination of market rigging, sweetheart deals with captive suppliers, and bold-faced conspiracies to kill off smaller rivals and to suppress all alternative [transportation] technologies, the Big Three [General Motors, Ford, and Chrysler] colluded to surf the marketplace as if it were their own private tsunami of profit. 'What's good for General Motors is good for the country' was more than just a slogan of the time. It was the rationale for an auto-dominated social regime whose effects were in many respects certainly *not* good for the country.[7]

For good or ill, though, a cartel is simply a set of *choices* applied to a social, economic, or political environment. It represents a unique balance between economic freedoms and restraint. This balance in turn reflects the common priorities of those who strike it, irrespective of whether their cause is public or private in scope. Consider government rules on the wholesomeness of food, the airworthiness of commercial jets, the safety of blood-transfusion products, and allowable emission of pollutants from nuclear power plants and from automobiles. These, too, are cartels of a sort. Public authorities have created a cartelized or 'unfree market' so that citizens can eat in public places, submit to hospital treatment, and travel to their destinations with an ease of mind that a wholly unregulated market would never permit. Tax and income redistribution are likewise designed to mitigate the worst excesses of competition. These measures have a cumulative effect. Although they are often taken for granted, individually criticized as being 'inefficient,' and rarely (if ever) measured in any meaningful statistical way, they combine to create that elusive atmosphere of continuity and confidence on which the whole of our prosperity rests.*

Likewise with computer software – another set of ordering principles. Operating systems such as Unix, Windows, or Mac are nothing but ways of managing information to achieve specific objectives – be they commercial, social, or political. Unlike printing presses, which also projected particular dreams and objectives, these new systems are inextricably woven into the very fabric of life, and facilitate practically every element of work and play in a fully wired society. Indeed, if the *polis* can now be usefully compared to a cybernetic system, then it becomes important to ask, once again, Who is *responsible* for its coding? Its operation? What are the rules of the game? Indeed, what priorities does the system pursue? While illuminati of the digital world suggest that complex systems are best left to regulate themselves, they mask the practical result, which is that business would be left to write its own rules. This would ensure that public 'input' in setting of standards would be marginalized and that the specific goals of enterprise would be favored at the expense of all else.

It is inconceivable that the economic (and political) rewards of such 'market anarchy' would be spread equitably (and automatically) to the public at large. To students of Reaganomics, indeed, it sounds suspiciously like 1980s–vintage 'trickle-

*This continuity is the result of societies' agreeing to sacrifice a degree of economic liberty in exchange for a measure of calm. Nowadays, as society splinters more and more according to socioeconomic lines, it is fashionable among the well-to-do to speak of such social costs as being 'inefficient and inflexible.' Many European economies are said to be 'overregulated.' Policymakers are asked to weaken standards so that national industries can compete more effectively in a 'global marketplace.' Yet many of the very same regulatory restraints helped to give Europe a level of relative social equilibrium which, even now, is the envy of many countries in Asia and the Americas. Social welfare schemes now being diluted throughout the developed world *did* prevent the recessions of the 1970s and 1980s from degenerating into a 1930s-style depression. An economy organized along more libertarian lines will be unable to do the same. Ironically, there is no evidence that high government social spending in any way constricts a country's ability to grow economically (see Peter Lindert, 'What limits social spending?', *Explorations in Economic History*, January 1996). But, because this myth has taken hold, the net result is that the poorest (who tend not to vote) are being further disadvantaged in relation to the middle class and wealthy (who do). It remains to be seen how long a state so organized can be sustained.

down' theory – dusted off, shrink-wrapped, and stamped with a digital seal. It conceals a momentous new political arrangement whereby the principle of 'one person – one vote' is to be replaced, by technical imperative, by the tyranny of 'one *dollar* – one vote.' This is why, as David Marquand provocatively suggests, 'either democracy has to be tamed for the sake of the market, or the market has to be tamed for the sake of democracy ... Globally and nationally, we shall sooner or later have to choose between the free market and a free society.'[8]

Imagine yourself in the year 2016. By then you will inhabit a world in which immersive virtual reality has become commonplace. To hear the digerati tell it, you will be able to slip into a 'data suit' and catapult into convincing three-dimensional environments for work and play. These will be programed worlds: illusionary landscapes, digital cities. The 'streets' may well be peopled by humanoid forms, avatars, software 'representatives' that we real-life participants have programmed to display the postures, motion-styles, and moods we wish to project. Conventions will govern whether (and how) people choose to 'interact' in this networked space. Walking along the streets of this virtual city, you might see a futuristic pyramid clad in coded gold, a tall cylindrically shaped structure in lapis blue, a third skyscraper in flaming red, pocked with 'windows' of bronze.

Just as companies erect skyscrapers today, the information powerhouses of the twenty-first century will fabricate algorithmic 'presences' on the Net. But a nagging voice tells you that these companies will play a radically different role than those of today. Although they will almost surely display many features of late-twentieth-century enterprise, they may also have captured many powers now associated with sovereign government, and thus have sidestepped the democratic procedures designed to limit autocratic excess. If so, such companies will be in a position to condition the lifestyles and opportunities for vast wired populations, on a scale and in ways that have never been possible before. They will dominate the scarcest information commodities and straddle the electronic routes along which they ride. A distant elite in cloud-capped towers of informational rule will design and powerfully condition the realities of the networked world. The biggest of these electronic domains may have their own security forces and judicial mechanisms, and even mint their own virtual 'coin.' Citizenship will be extinct. Communities will have disengaged from history and physical geography. Uprooted, fragmentary, and ethereal, they will take shape and dissolve as fast as electronic ties can be set up and pulled down. Today's techno-futurists predict a world of 'consumer communities' in which life, for our descendants, will involve navigating through a floating world of code.

How did we get here? Reel back, if you will, to the year 1990. Edzard Reuter, chairman of the German technology giant Daimler-Benz, is being piloted to Paris aboard his executive jet. With him are only his personal secretary, a member of his PR staff, and a London-based journalist. As we punch up through a layer of particulate smog, a reddening autumn sun lights a garishly strange cloudscape to the West. Reuter looks momentarily disturbed, and frowns. Fitfully, he sips at his tea. Then, leaning into a black leather chair, he signals to the reporter that he is ready to begin.

The tape begins to roll.

Reuter has just wrapped up an important industrial alliance with counterparts at Mitsubishi in Tokyo. Now, high above the clouds, he lays out his vision of the industrial path ahead. The greatest and most momentous enterprise, he says, is to achieve 'a networking and interdependence of big business interests all over the world, so that no politician will ever be able to disrupt it without risking his own viability.'[9] *No disruption?* The journalist is alarmed. In fact the ambition of Reuter (who has since retired) remains the ambition of many corporate leaders today. Akio Morita, who founded the Sony Corporation, has called for a completely 'harmonized world business system with agreed rules and procedures that transcend national bounds': one in which, by clear implication, local relativities and political procedures would give way to the imperatives of wired global enterprise.[10] Across the Atlantic, a New York-based telecommunications executive speaks in virtually identical terms, celebrating the fact that nation-states are already being transformed into what he calls *virtual lobsters* (that is, creatures on which it is the digerati's pleasure to dine). As he sees it, 'The shells of these sovereign crustaceans have ... proved too porous to prevent their contents from being cooked to someone else's taste.'*

Whether in Europe, Asia, or the Americas: the message is the same. Imposing too much accountability on wired enterprise would stifle competitive growth. Government should either cooperate or get out of the way. By and large, government plays the game.† The elites of the public and private sectors now enjoy a more relaxed and conciliatory relationship with each other than they have in years: a case of power following money. In a more complex world, politically sensitive issues can now be laid at the table of multilateral organizations. These are conveniently staffed by unelected bureaucrats who understand both the rewards associated with toeing the party line and the penalties for breaking ranks. Powerful institutions like the IMF, the World Bank, the newly formed World Trade Organization (WTO), and the European Council of Ministers conduct their daily operations behind impenetrable cowls of monastic secrecy.‡ The procedures, as one ranking EU official concedes, 'are utterly undemocratic.'[11] Essential issues affecting the health and welfare of the

*Frederick S. Tipson, 'Global communications and the sovereignty of states', p. 2 (see the footnote on page 138). The great irony is how many 'global' companies still wrap themselves in national flags to obtain tax breaks, regulatory concessions, and subsidies from their 'home' governments.

†For example, research into the impact of electromagnetic microwaves from mobile telephones on the human brain has fallen far behind the growth of the market itself, which is increasing at a rate of 50 percent a year (and reached 85 million subscribers in 1995). Business and regulators grudgingly concede that further study is required to definitively establish whether such phones cause brain cancer and other conditions ranging from asthma to chronic nausea, but they are, in the interval, unwilling to withdraw from the scramble to win market share.

‡The WTO recently launched an 'openness' campaign to 'de-restrict' its documents – all, that is, except those contained in all-important 'Appendices' which happen to include agendas, working papers, decisions, and proposals. Thus, essential information about decisions affecting all areas of public life is unavailable to citizens in time for them to react. Consumers International, a federation of more than 200 consumer organizations worldwide, has warned that 'the veil of secrecy surrounding the WTO compounds public anxiety about supranational rule and discredits the international trading system. It perpetuates the perception that trade agreements are dictated by a few behind closed doors.'

world's population are being decided in opaque conclave by an 'overlay culture' that is constitutionally unaccountable to the wider human constituency it serves.* This helps explain why it was possible to establish an organization to manage global commerce – the WTO in Geneva – without a countervailing institution and/or procedures to govern the overall ecological and social context in which the trade ought to proceed.

Thus, what might have seemed like dystopian science fiction a few decades ago is now, disquietingly, all too real. As the civic space of nations is effectively migrating into the electronic space of the Web, its rules of association – democratic or otherwise – have yet to be defined. John Garrett, at the Corporation for National Research Initiatives (CNRI) near Washington DC, is

*Given the recent scandal in connection with the public sale of BSE-tainted beef, it is interesting to consider the circumstances under which the Codex Alimentarius commission met in its last, nineteenth, session to rework food safety standards in light of the globalization of markets. The meeting was powerfully dominated by chemical- and food-industry representatives. Some 2,500 people attended. Of these, 660 were industry representatives who had come to 'assist' the national delegations. There were 105 national delegations, and 140 multinationals, including the food groups Nestlé, Unilever, Coca-Cola, and the Philip Morris subsidiary Kraft. Nestlé, of infant formula-milk notoriety, sent more delegates than most sovereign states. Indeed, at the additives and contaminants meeting, over 40 percent of the participants were industry representatives. Set against these 'advisers' were twenty-six men and women representing the world's public-interest non-governmental organizations (NGOs.) In the final agreement, it was allowed that costly standards governing such matters as the levels of animal growth hormone, carcinogens, and radiation levels in food could be lowered 'at national discretion' to accommodate the interests of a 'fair and competitive playing field' for the industries concerned. (Figures from *Cracking the Codex*, UK National Food Alliance, 1994, and other sources.)

When the G7 industrialized nations met in Brussels in February 1995 to discuss construction of the 'information superhighway,' for example, 'the hidden agenda was the speed at which countries ... were prepared to liberalize their telecom regimes,' writes Alan Cane ('First steps towards a structure for global communications,' *Financial Times*, 27 February 1995). 'It was the first time industrialists had been invited to share in the proceedings of a G7 conference.' Non-governmental organizations, a few of which are in a position to articulate public concerns on a global scale, have minimal 'input' in a process whose importance has flown over most people's heads.

The dissipation and fragmentation of public responsibility was likewise blamed in the case of Germany's 1994 blood scandal, when it was discovered that many hospitalized Europeans had been exposed to HIV-tainted blood products supplied by a financially ailing medical-supply firm. These and many other cases expose the vulnerability of an underregulated and increasingly networked market to full-blown systemic shocks as well as to abuse by weak and/or rogue operators.

part of a cross-industry working team that is formulating common standards for America's 'information superhighway.' He is convinced that the advent of networked enterprise will bring us nearer to 'a world where there are going to be lots of dynamic relationships – exceedingly profitable within a very short time frame – but cross-border, cross-jurisdictional, cross-everything.' When asked about a new set of democratic procedures to provide for a regulatory counterbalance, he throws up his hands and asks, 'Rules? I just don't see how you'd design them. After all, how are you going to measure market dominance when it lasts thirty seconds?'[12]

Booz-Allen & Hamilton's Cyrus Freidheim takes a somewhat different view. As he sees it, regulatory power will shift further toward networked industry. There will be many small and dynamic companies and independent entrepreneurs who live 'portfolio' lives, but the future economy will be primarily dominated by a species of communications-based relationship enterprise that he has called the 'Trillion Dollar Corporation.' Although such a company's *products* will be more finely targeted – and therefore have shorter market lifecycles as John Garrett suggests – the underlying business *relationships* will grow progressively more strategic in scale and scope. What's more, because such companies will be shaped and driven by the nature of their *electronic interconnections*, as opposed to traditional (and visible) cross-shareholding ties, their activities will be more opaque than those of industry today. To illustrate his point, he offers the following scenario, which is only just quasi-imaginary.

AT&T, Sony, Time Warner, Motorola, Nynex, and Japan's KDD join together to develop a new interactive communications and entertainment network. They begin by leveraging their combined political clout to sweep aside regulatory obstacles to a common technical standard.* Then, they jointly invest billions in construction of the delivery system.† After an interval, they have the network up and running. This enables them to spin off derivative products and services, to build up a

*For example, Visa International/Microsoft and MasterCard International/Netscape have joined forces to collaborate on a new security system for credit transactions over the Internet: 'the single standard limits unnecessary ↪

costs and builds the business case for doing business on the Internet' (Louise Kehoe, 'Credit card groups to co-operate on Internet security,' *Financial Times*, 2 February 1996).

†In fact six of the world's leading telecom companies took the first steps toward such a system in 1993, when they began construction of a worldwide fiber-optic submarine cable network dubbed the 'Global Networking Project.'

profitable momentum, and thus attract still other large corporations into their fold. What all began innocently enough as a specific, project-oriented alliance suddenly seems to be widening and maturing into a different beast altogether. The partners are learning to trust each other; they are finding areas of mutual interest; they are developing a common agenda. Before long, they are acting together as if they were a single, coherent enterprise: one great galaxy revolving around a solar hub. The relationship enterprise takes on attributes of a sovereign state.‡ Its true strength probably cannot be gauged by the formal cross-shareholdings between the various partners. It is more invisibly embedded in electronic interconnections: in the lines of communication that facilitate the quickening process of delivering ever more finely targeted products to market in shorter periods of time. As Freidheim sees it, 'An enterprise of this size, acting in unison, has the potential to *shape* the environment, rather than respond to it . . . it would have the resources to pursue initiatives that no single corporation could undertake . . . it also could, to some extent, sidestep antitrust . . . '13

‡Among the powers usually reserved for sovereign nations are the management of the financial system, maintaining a judiciary with police and penal powers, and supporting an administration to form policies, a civil service to carry out laws, and a military to assure independence from external threat. Already the multinational enterprise 'encroaches on areas over which sovereignty and responsibilities have traditionally been reserved for national governments,' reports the United Nations (in *World Investment Report 1993: Transnational Corporations and Integrated International Production* (New York: United Nations Publications, 1993), cited in Tony Jackson, 'Multinationals take lead as world economic force,' *Financial Times*, 21 July 1993). This trend will almost certainly be amplified in networked space.

In the absence of sufficient procedures for democratic oversight, relationship enterprise will drive public–health, environmental, and workplace-safety standards down to a least common denominator. And evade paying a full share of tax. And build dossiers on the private lives of individuals for commercial gain. This version of cybernetic 'freedom' is one in which businesses are disengaged from the legitimizing soil of the communities within which they operate. After all, as *Business Week* points out, 'modern multinationals are not social institutions. They will play governments off against one another, shift pricing to minimize taxes, seek to sway public opinion, export jobs, or withhold technology to maintain a competitive edge.'[14] *The Economist* notes that 'in the ... de-ideologized politics of post-Communist days, the lobbyist is getting ever more powerful ... and democrats are right to be worried.'[15]

This is precisely what disturbs a circle of long-term thinkers who find themselves marginalized, at least for the moment, from the current debate. One senior Clinton-administration official in Washington – an experienced public servant who has grown so cautious that he refuses to discuss the emerging informational monoliths except on terms of strictest confidence – phrases it like this: 'You can already see how they carve up and control markets. It is simply a matter of their gradually developing more control, and more efficiency, and ... leaving many of us to jump up and down exclaiming, "Free market competition! National champions!" And you know what? We'll be missing the entire point.'*

The point is this: just as sovereign currency threatens to be eroded by the arrival of digital cash, national political institutions are gradually being bypassed by insufficiently accountable *free states*, or information-based domains.

Bounce this across to John Perry Barlow and he hits it back with a spin: this is *exactly* the point. 'Today,' he pronounces, 'centralized

*Background conversation with senior Clinton-administration official, Washington DC, 21 July 1994. The official points to several weaknesses in the model of an entirely free-form networked economy. First, in a laissez-faire world, big informational powers will not voluntarily interconnect with each other on equitable terms unless they perceive it to be in their own interests. These, in turn, will not necessarily (or often) coincide with those of the general public. (A good historical example is the telephone network: the ability to call anyone, regardless of which network

carries the signal, is the result of a mandated sharing of costs and benefits. Nowadays new networks are evolving in which the proprietariness is embodied in the way they are programmed to work. For instance, an airline reservation system might contain all the data a person could want. But information about Cartel X's flights, cars, and hotels is much easier to access than data on deals offered by the rival Cartel Z.) Second, the market power of globally ↪

networked relationship enterprises will be so strong that it may work against a vital and competitive economy. Given the bargaining leverage provided by massive economies of scale and scope, the networked enterprise will be in a superior bargaining position relative to smaller

knowledge vendors, all of whom have only a narrow view on the overall galaxy. Third, the market cannot solve fundamental distributional questions – including, but not limited to, tax policy. Nor can it be relied upon to safeguard the public good in such areas as standards for environmental protection. While the mechanisms of the market can be useful tools in all of these areas, they can be so only when counterbalanced by political means.

government serves no useful purpose at all.' Indeed, 'the only thing it really does well is get in the way.' It has no legitimate authority over the informational realm. As the world grows progressively more wired, cyberspace is evolving into 'a sovereign entity unto itself.' It is a force for creative change in a world of obsolescent ideas. All of which may seem persuasive, up to a point, when applied to the Internet alone. The trouble is, it does not apply to the whole of information space. Cyberspace is not a *single* integrated and ubiquitous entity revolving around the Internet. It is a place fractured into *multiple spheres of influence*, each with autonomous priorities of its own. Sometimes these spheres will cooperate, sometimes they will compete, but always they will guard their vital cores.[16]

Abbe Mowshowitz, an accomplished Canadian scholar, looks around the contemporary landscape – supposedly unlike any that ever came before – and sees clear historical parallels. While digital 'revolutionaries' herald an unprecedented era in human affairs, Mowshowitz sees new information hubs emerging as something comparable to feudal estates.* He is also convinced that today's general institutional decay is powerfully reminiscent of the closing days of the Roman Empire and the spread

*This is a world in which prison management, tax collection, and policing become privately run. In the US, classrooms have already become profitable outlets for corporate advertising; worldwide, there are growing calls for a complete privatization of public school systems. At the start of 1995 there were more people involved in providing security on behalf of private-sector firms in the UK than there were employed by the publicly funded police. Large areas of the public sector, including the collection of taxes, have now been contracted out to private companies like EDS, the US information-management group. In the US, some judicial procedures and many prisons have also been privatized – a phenomenon that happens to coincide with the rise of a successful new entertainment network, Court TV.

of the so-called Dark Ages. On the economic front, he writes, the emerging relationship enterprises:

> ... are likely to turn into the fiefs of virtual feudalism. Instead of depending, for the creation of real wealth, on tenant farmers bound to the land, the emerging order will depend on production units bound to global firms operating as virtual organizations. De facto political authority will shift to the virtual fief because government, as we know it, will lack the money to maintain many of the judicial and social services currently provided. No doubt governments will play some role in virtual feudalism, since social institutions rarely disappear completely; but their role will be diminished and probably be confined to providing local protection of infrastructure ... [17]

Neal Stephenson, a novelist who chronicles the millennial force-field that is wired society, paints a similar picture in a slyly 'futuristic' work called *The Diamond Age*. The book describes a world of Equity Lords and economic city-states that are regulated by a Common Economic Protocol (only a modest extrapolation from the free-trade agreements guiding the global economy of our day).[18] In such a networked landscape, the scope of creativity and cultural expression will be powerfully bounded by the priorities of informational overlords, whose power as *signal managers* will actually be enhanced by the sea of static they create. According to Esther Dyson – a consultant who is one of the most powerful of the few woman in computing – 'just as prominent patrons such as the Medici sponsored artists during the Renaissance, corporations and the odd rich person will sponsor artists and entertainers in the new era. The goal for advertisers,' she states, 'is to make [sure] ... their advertising messages are ... inextricable from the content that surrounds them.' The ultimate intent is 'to control ... a relationship with the customers.'[19]

The key word is 'control'. When people gather in churches and civic crossroads – physical spaces in which differences get *bridged* – they also enjoy an integrated social experience that forms both a source of inner strength and (where necessary) a moral foundation for effective political resistance. On the Net, such experiences can be progressively marginalized. Personalities and entire cultures can be systematically isolated and carefully dismantled.

Marshall McLuhan foresaw this danger in his now-famous 1969 interview with *Playboy* magazine. He explicitly warned that the arrival of an ambient cybernetic web will make it feasible, for the first time, to pattern the senses of society through a 'total media experience.'*

> Once we have surrendered our senses and nervous systems to the private manipulation of those who would try to benefit from taking a lease on our eyes and ears and nerves, we don't really have any rights left. Leasing our eyes and ears and nerves to commercial interests is like handing over the common speech to the private corporation, or like giving the earth's atmosphere to a company as a monopoly.[20]

*McLuhan spoke of using computers to 'conduct carefully orchestrated programming of the sensory life of whole populations ... [and to] determine the given messages a people should hear in terms of their overall needs' (The Playboy interview, *Playboy*, March 1969).

There is nothing new about the goal or method involved, only the tools that are being being used. As the Kenyan artist Ngugi wa Thiong'o remarks, in his classic study on the modes of eighteenth-century imperial rule, 'the most important area of domination was the mental universe of the colonized; the control, through culture, of how people perceived of themselves and their relationship to the world. Economic and political control can never be complete or effective without mental control.' Indeed, Ngugi wa Thiong'o adds, 'to control a people's culture is to control their tools of self-definition in relationship to others.'[21]

Esther Dyson is mapping a future based on exactly this view. An important intellectual force behind the much-celebrated *Magna Carta for the Knowledge Age*,[22] she is one of the few women who can boast of accomplishments and impeccable connections at the highest levels of the amorphous new elite. She has synthesized a remarkable idea. She believes that what we now call the Net will soon be carved into a series of 'customer communities.' They will revolve around corporate consortia that include airlines, supermarkets, phone networks, privatized utilities, and the rest. And, since each hub may well sport its own electronic chip currency, this will gradually give rise to a new state of economic and political loyalties. As a customer, you will be granted 'user credits' whenever you make a purchase. These will be like frequent-flyer miles, except that they will be redeemable for a much wider range of goods and services, including those that you need to survive.[23] Thus, buying a particular good or service will both identify you and reward you for loyalty. (In fact the act of buying will be fundamental to the establishment of personal identity.) Unlikely as it may now seem, some people will make a livelihood collecting such credits: their earnings will be derived merely from identifying themselves and participating in cyberspace manifestations on the Net. As Microsoft's Bill Gates has remarked:

> [The Net] will be able to sort consumers according to much finer individual distinctions, and to deliver each a different stream of advertising ... [indeed] some of the billions of dollars now spent annually on media advertising ... will instead be divvied up among consumers who agree to watch or read ads sent directly to them as messages ... there would be a group of 1-cent messages, a group of 10-cent messages, and so forth. If there was no fee attached, they could be refused ... but if someone sent you a $10 message, you would probably take a look ... *[24]

What's more, because so many aspects of consumers' life will be mediated by the Net, Gates suggests, a sort of 'friction-free capitalism' will emerge. It will be possible for 'those who produce goods to see ... what buyers want, and will allow potential consumers to buy those goods more efficiently. Adam Smith would be pleased.'[25]

Will this vision prove sufficiently efficient or workable to justify the

*In the fall of 1996, CyberGold, a Berkeley, California-based company backed by Regis McKenna (a leading Silicon Valley PR strategist) and Jay Chiat (co-founder of the Chiat/Day advertising agency), inaugurated a service offering e-mail users the opportunity to sign up as willing recipients of junk e-mail in exchange for a payment of at least 50 cents per message. Users can also sell their terrestrial address to advertisers for $8. They receive their payments in the form of electronic *tokens* that can be redeemed for a range of products and services.

substantial strategic investments being made in its name?* A society is built on shared obligations, on compromise, but what incentives are there to compromise in a digitized social landscape of two-way telescreens, total sensory immersion, and 3D video holography – one that offers precisely as many possible experiences and opportunities as there are people tethered to the Net? In these realms, as MIT's Nicholas Negroponte remarks with chilling enthusiasm, each of us will be an audience and a market of one.[26] Our morning newspapers – and our vision of reality – will be composed of data that have been especially filtered to appeal to our documented taste.† The tribal space as a whole will be reconfigured into a pandemonium of electronic 'interactions.' As an added bonus, space-time will be fragmented as well. For those souls who wish to communicate with each other, above all, there will be *costs* associated with the privilege. True connectivity will become a function not of openness and/or persuasion but rather of an ability to pay.‡ A world nominally dedicated to satisfying the individual will at the same time be one in which individuals have been largely inhabited and *colonized* by the signal managers of the information space. They will truly have become – as Jean Baudrillard once said – *fractal subjects*.[27]

The operative idea seems to be that a radically market-oriented society of fragmented consumers, each seeking to maximize his or her own individual gain within the Web, will produce not only a vital economy but also, somehow, a

*Certainly, 'consumer communities' will be very different from living cultures as we understand them today. They will be defined more by *purchasing* choice than by the wider complex of shared inclinations and implicit beliefs. Consumer communities are predicated on the movement of messages. Civil communities must be capable of truly communicating, of compromising, and of accepting unprogrammed sacrifices in pursuit of some higher shared ideal. They depend on an extraordinary (if fragile) set of democratic procedures that are *constitutionally* delineated and have equal application to all.

†Chapter 3 describes how the introduction of cybernetic systems in business has triggered a revolution of 'mass customization.' Where the world of mass production would produce a single Ford Model-T available to all in black, companies can now wheel out an astonishing array of goods and services, with ever-shorter lifecycles, tailored to more narrowly defined markets and ever-shifting trends in taste.

‡Bill Gates suggests, 'if someone not on any of your lists wants to get to you, he'll have to have someone who *is* listed forward the message to you' (Bill Gates with Nathan Myhrvold and Peter Rinearson, *The Road Ahead* (London: Viking, 1995), pp. 213–14).

sustainable culture and a healthy environment. There is no need to seek a happy median that combines the creative energies of chaos with the conservative restraints of an enlightened civil administration. Indeed, given the present frustration with the status quo, and the toxic excess of a corrupt and regressive political establishment that stands seemingly powerless at the cusp of a new millennium, perhaps it comes as no surprise that pent-up demand for such radical change – indeed *any* change – exists. So, an interval of chaos may well be on the cards: must citizens wait until its smoking aftermath before they start to ask, What next? By then there is no guarantee that democracy itself can be dragged out and revived from the rubble of elective government that remains.

Meanwhile, the process of 'mass customization' is already well under way, and its fundamentally economic motive is routinely ignored in public proclamations about the tantalizing road ahead. What the digerati are in fact engineering is a wholesale colonization, privatization, and commercialization of the human sensorium. As the president emeritus of AT&T's Bell Labs, John Mayo, suggests, the so-called 'decentralization' brought about by information technology is at root a remarkable marketing opportunity – at least for those who are most capable of exploiting it. He notes that 'the technology is increasingly able to customize right down to the smallest buying unit you can find, which is the single individual, single student, single member of society.'[28]

Note the signification implicit in that code.

Small. Single. *Buying unit . . .*

Many digerati, like John Barlow, consider this train of thought misplaced and absurd. It will be *people* who configure the Net, he and others insist. And, when experience is transformed into a sum of personal desires, then 'users' will once again have become 'co-authors' of their own destinies – just as Australia's Aborigines authored their life experiences in 'the Dreamtime.' But the sounding-board of Aboriginal identity was the collective experience of moving through a shared geographical space. The natural landscape was fixed, and each individual had unwavering responsibilities within this mutual terrain. Individuals could hardly *choose* the disciplines that nature imposed: they could merely hope to *negotiate* them at best. It was this process of negotiation – of continually *communicating* with one's physical surroundings – that cultivated a preference for healthy rather than dysfunctional relationships.

Nomads of the digital world, their heads metaphorically and literally nested in the glorious isolation of head-mounted video displays, will evolve in different ways altogether. Their likes, dislikes, and affiliations will be shaped primarily by their *personally selected* interactions within an infinitely fragmented electronic space. (The great Italian playwright Luigi Pirandello anticipated such freedom a century ago when he wrote that 'Life is a very sad piece of buffoonery, because we have ... the need to fool ourselves continuously by the spontaneous creation of a reality [one for each and never the same for everyone] which, from time to time, reveals itself to be vain and illusory.'[29])

Unless citizens of the electronic age can focus more sharply on the commonalities that bind them – and overcome the fragmentation of their existential space – and unless it is possible to configure systems of democratic *accountability* that are equal to the gregarious sprawl of the networked world, then all of the high hopes for an electronic 'commonwealth' will be dashed. Instead, citizens may well find themselves inextricably wired into a Panopticon prison – one where the 'walls [are] so high that those inside cannot see each other.' Here, in this unprecedented regime of remote control, only the Webmasters will enjoy a God's-eye view.*

*The quote is from Jeremy Bentham's classic treatise *Panopticon*, as given by Michel Foucault in *Discipline and Punish: The Birth of the Prison* (London: Penguin, 1987). 'The volume of requests and transactions likely to be funneled through the head end system would probably require an extensive interconnected network of local or regional mega-processing facilities housing enormous communications, computing, and video broadcast equipment and services' (Goldman

In 1776 Adam Smith's now-classic work *The Wealth of Nations* was published. Smith realized that ageless and seemingly immutable certainties were being challenged by the Industrial Revolution, and he sought to fashion a fresh conceptual framework within which people could understand the great changes that were sweeping through their lives. He hit upon the useful notion of comparing the common-wealth to an industrial machine. Before long, people started talking in mechanistic terms – wielding *levers* of power and, as often as not, feeling like *cogs in a wheel*. How-ever, the fact that people could not fully conceive of themselves in this

Sachs Investment Research, *Communicopia: A Digital Communications Bounty*, July 1992, p. 61). When aspects of sensory stimulus are managed by the centralized servers and soft-ware routines, those at the crossroads of the web, coordinating, collating, and filtering *signal* from vastly increased noise, will be most empowered. The scarcity value of essential information rises as the body of raw data is increased.

mechanical mold of early industrialization, nor initially organize themselves in an equitable way, were factors in the violent and instinctive backlash against it. As Karl Polanyi wrote, 'to make Adam Smith's "simple and natural liberty" compatible with the needs of a human society was a most complicated affair.' It required 'no less a transformation than that of the natural and human substance of society into commodities ... [and] the dislocation caused by such devices ... [was to] disjoint man's relationships and threaten his natural habitat with annihilation.'[30] The tensions were eased with the creation of the social welfare state, which moderated the excesses of transformation and actually represented an inspired modern adaptation of value systems evolved by many indigenous cultures centuries before.

Now, well over two hundred years later, amid intoxicating doses of speed, mobility, and abstraction, and with its economic lifeblood pumping through a sprawling web, society faces a similar upheaval in outlooks and institutions. In response, the digerati, 'feigning mystic glamor and lordly attitudes,'[31] propose that it is best to kick back and let the whole show proceed according to some 'default parameter' in the system itself. However, standing on his vantage point at the dawn of the electronic age, Norbert Wiener, the father of cybernetics, looked presciently ahead and wrote the following words. 'The future offers very little hope for those who expect that our new mechanical slaves will offer us a world in which we may rest from thinking. Help us they may, but at the cost of supreme demands upon our honesty and our intelligence, not a comfortable hammock in which we can lie down and be waited upon by our robot slaves.'[32]

In the early phases of almost any great transformation, extremes of utopian optimism usually prevail at the expense of common sense. When the telephone first arrived on the scene, its exponents insisted that it would act as a promoter of world peace, a savior of the lonely, and a channel for enlightenment against 'heathenism' in all lands.[33] Alexander Graham Bell, who patented the phone in 1876, was convinced that its primary use would be to broadcast classical music concerts and Shakespeare plays. In fact, this new communications infrastructure, which complemented the railroad and telegraphy, gave rise to powerful empires and new models of economic, social, and political life which, in turn, posed crucial questions of human equity. The markets, then as now, proved imperfect and anti-competitive. Elected officials were obliged to step in.

Now citizens are told that a globally networked business environment will render government obsolete. The truth is exactly the reverse. For all the loose talk about globalism, those very companies that are now morphing into 'relationship enterprises' still nevertheless generate about three-quarters of their value-added on their own home turf.[34] Moreover, they still depend crucially on close ties with government, stable industrial relations, and dependable sources of physical supplies. According to one important study, the reality behind the myth of globalization is that it 'provides big companies with a major bargaining chip in their negotiations with suppliers, organized labor and governments.'[35] It also makes it easier for politicians to evade responsibility for their deregulatory policies by allowing them to cite the abstract imperatives of 'global competitiveness.' (Ironically, 'at least 20 companies in the 1993 Fortune top 100 would not have survived as independent companies if they had not been saved in some way by their governments.'[36])

In fact it is precisely in the midst of a chaotic technological transformation that ancient choices are posed in new ways. The issues of 'free' versus 'deregulated' markets, of pornography and Internet censorship versus total freedom of speech, of government attempts to limit the spread of advanced encryption tools – all boil down to profound and timeless dilemmas. The underlying question remains: Do citizens have the right to place limits on absolute freedom – in what cases, and toward what goals? What is the correct balance between complete openness – for example in trade and cultural relations – and partial restraint? What is the best way to promote pluralism and the open exchange of ideas? These tensions remain essentially the same, even if the social space is being radically changed. As markets migrate into a unified digital space, and as the marbled pillars of traditional democracy confront the quicksilver *free states* of the binary world, the essential challenges are to configure a new system of political checks and balances, to equalize access to available opportunities, and to make sure that these priorities are pursued within a socially accountable regime.

There are never ideal solutions, only less imperfect ones – and these can only be legitimately decided by citizens and their credible representatives. The 'freedom' promised by the digerati is one of radical social distortion and political tyranny, a life of unrestrained social chaos that is nothing but 'nasty, brutish and short,' as Thomas Hobbes famously remarked. Moreover, such a world would be one that is increasingly devoid of the arts, letters, and civilized refinements; it would enter, as Hobbes said, 'a condition of war.'

One possible method of reviving engagement in the affairs of modern democracy – and of reducing the unhealthy influence of self-interested professionals and monied special interests – is a 'civic service obligation.' This would require every adult, as a condition of citizenship and for a fixed number of years in their lifetime, to work in and help manage their government in a capacity appropriate to their abilities, education, and life attainments. Meanwhile it is time that digital free-marketeers take in a message themselves: those who wish to profit from operating in a shared space must also accept the inherent responsibilities. Market stability, public order, an educated citizenry, and indeed the very future of society itself, depend on everyone pulling their fair share.

9
Sandcastles

If you are playing a game according to certain rules and set the playing-machine to play for victory, you will get victory if you get anything at all ... the machine will not pay the slightest attention to any consideration except victory according to the rules ... [and] at any cost, even that of the extermination of your own side, unless ... survival is explicitly contained in the ... [way] you program the machine.

NORBERT WIENER[1]

When he died in 1995, France's Louis Malle ranked among the most accomplished, if exasperating, figures in the world of film. Among his most highly acclaimed and controversial works was a 1968 documentary on India called *L'Inde Fantôme* (also sometimes known as *Calcutta*). In one scene, the audience is shown the denouement of a rural train disaster. The scene begins with a close-up shot of what seems to be a pile of hopelessly battered iron. Then the camera pulls back, the scene opens out, and we realize that this gnarled black mess is an intact but derailed steam engine. It has come to rest, improbably upturned and inert, beside the rails. Malle's lens swings slowly up the embankment. The screen fills with the countless passengers and villagers standing by. Then, with mounting disbelief, we watch as a thousand palms close upon the steel and start a coordinated push. Inch by painstaking inch, using nothing more than the power of their muscles and the simplest of wooden slats for leverage, this determined congregation literally heave the train back on its tracks. It is a remarkable image of concentration, of living energy, and of the sheer power of the human will. Watching this scene can set you musing for days. Will the metaphorical train of high-tech society be similarly salvageable if it ever runs off its rails?

It is lunchtime in late January 1990. A technician at AT&T's network control center in New Jersey is sipping a cup of coffee when something strange flickers at the edge of his screen. It's there again – an odd sort of glint   like a subtle flutter through the main-sail before a hurricane closes in. Glancing up from his screen to the national control grid, a series of several hundred computer displays tiled across the far wall, the technician looks on in horror as they project a tide of red warning lights suddenly sweeping across the face of North America. Within seconds, AT&T's entire long-distance switching system crashes onto its knees. When the world's biggest computer crashes, the US economy is not far behind. The communications blackout of 1990 lasted nine hours, and it remains a legend among computer professionals today. They know all too well that it can – and will – happen again. The only question is: on how wide a scale?*

As systems become more complex, the difficulties associated with controlling them escalate too. These systems are fundamentally different from continuous natural and organic 'systems,' in which small changes or mistakes generally tend to produce an appropriately localized response. During the Allies' strategic bombing of ball-bearing factories in the Second World War, for example, the Germans arranged a general redundancy of factories and supplies that let them resist much longer than the tonnage of dropped ordnance suggested was possible. When gears and levers gave way, mechanics rolled up their sleeves, replaced individual components where necessary, and generally put things right. (Indeed, the Third Reich shared at least *this* with its

*On 7 August 1996 America Online's computer systems went down for nineteen hours, affecting 16 million subscribers. The crash was caused by new software installed during a scheduled maintenance update. 'It's another reminder of how close to the edge a lot of these systems are,' said Richard Zwetchkenbaum of International Data Corp. According to David Einstein, in an article published in the *San Francisco Chronicle* the following day, AOL officials had been boasting about the fail-safe integrity of their system. Indeed, one representative told the *Chronicle* that AOL computers were 'virtually immune' to this kind of outage.

enemies: it had an analog economy that showed resilience under punishment received!) In a cybernetic environment things are very different indeed. All it takes is a modest anomaly in a binary control string to bring an entire system to its knees.* For instance, the AT&T collapse was caused by the failure of one logic circuit that propagated from a single node through the whole of AT&T's vast long-line network.† Likewise, in June 1996, the first Ariane 5 rocket, developed at a cost of FFr 40 billion to maintain Europe's lead in positioning heavy satellite payloads into orbit, veered radically off course and had to be destroyed forty seconds after its launch from the Kourou space center in French Guiana. According to the inquiry board, its failure was caused by 'a complete loss of guidance and attitude information ... due to specification and design errors in the software of the inertial reference system.' The board cited inadequate analysis and testing of the rocket system's software design and called for a critical reappraisal of all the software and systems integration.‡ This caused a serious delay in further launches and raised

*On 17 July 1996 the Clinton administration called on US industry to mount the equivalent of a 'Manhattan Project' – the effort that produced the first atomic bomb during the Second World War – in order to create defensive safeguards for the vulnerable computer (and other) networks which have become the lifelines of modern society.

†Computer risks specialist Peter G. Neumann writes that 'the global slowdown was triggered accidentally by a crash of a single switch. However, a maliciously caused crash ... would have produced exactly the same results ... [and] illustrates the interrelations between reliability and security' (*Computer-Related Risks* (Reading, Mass.: Addison-Wesley/ACM Press, 1995), p. 127). Any discussion of cybernetic vulnerability owes a great deal to Peter Neumann. He has played a considerable (if initially solitary) role in raising general awareness about the subjects raised in this chapter, both as moderator of the Internet news group comp.risks and as a frequent contributor to *Communications of the ACM*, a respected journal for international computing professionals. For those wishing to pursue this subject, his book is the definitive reference work.

‡Jean-Marie Luton, director-general of the European Space Agency, at the ESA's press conference in Paris, 23 July 1996, and findings of the independent inquiry set up by Luton and the French space agency CNES. Evidently, the rocket's internal reference system was incapable of accommodating the unique flight trajectory of Ariane 5, which accelerates much faster than earlier versions of the rocket. 'I can identify seven chains of events which all contributed to the failure, and at each there were teams of people making decisions,' said Lennart Lubeck, a member of the Swedish Space Corporation who vice-chaired the board of inquiry. 'But it was inadequate foresight and an omission of logic. To fix the problem that caused the failure can be done in a matter of weeks. To make sure there are no other omissions takes months.'

doubts over Europe's leadership in the $3 billion-a-year space launch industry.

Such complex systems are no longer limited to high-tech space programs and the telephone net: they are embedded in critical applications all around us today.★ What's more, they are interconnected with other systems in vast geographically distributed webs. The whole of the world economy is being rewired as a great electronic grid. As a result, the same trends that are accelerating the speed at which we operate – and simul-

taneously insinuating cybernetic processes ever deeper into the fabric of modern life – are also exponentially magnifying the potential scale and scope of both human and systemic error. And note: although in uncertain times it is tempting to blame machines for failings that are more properly human at root – a tendency that often says as much about our own fear of the unknown as it does about any inherent weakness in the tools – the truth is that nowadays any local technical quirk can quickly escalate into a pounding global headache.

Because not even software engineers can predict the future, they tend to design against the likeliest risks. Tokyo, for instance, is elaborately geared for the ever-present danger of earthquakes. Its systems have been designed accordingly. At Kobe, in contrast, designers were more concerned with the challenges associated with typhoons. These are common events on a coast exposed at the turbulent confluence of the Pacific Ocean and the China Sea. When a devastating earthquake hit Kobe in 1995, there were no contingency plans to deal with such an unanticipated event: the region was assumed to be seismically inactive. The first tremors severed the fiber-optic cables on which the phone system relied and knocked out a large part of the city's communications. Physical roads, bridges, electricity supply, and transport were hit next. However, the most profound aftershocks were felt at far distant points around the planet: by car manufacturers and watchmakers deprived of crucial parts. A vast distributed network of interconnected 'nodes' – built around streamlined production and 'just-in-time' supply – was exposed as vulnerable to a combination of an unanticipated event, human error, and an insufficiently robust design.[2] Similarly, in July 1993, when a plant explosion ripped through Sumimoto Chemical's Ehime Works in Nihama, Japan, it took out the world's largest supplier of high-grade epoxy resins. As it happens, these esoteric resins are essential to the process of making semiconductor chips. During the five months it took for the plant to rebuild, rival manufacturers were at each other's throats over how to divide a 50 percent shortfall in supplies. Such is the stuff that wars are made of.

In Ecclesiastes it is written that there is nothing new under the sun. This seems more true for human nature than it does for our machines. The extraordinary pageant of technical innovation has been punctuated by explosive spurts of development in agriculture, construction, mining, medicine, printing, transport, and all the rest. But the *application* of these inventions has been more deliberate. Systems were adopted gradually, in a process of trial and error. This yielded resilience. Every advance built on those that came before. The overall structure could therefore withstand the trials of time and tide reasonably well. For example, new building techniques and materials were implemented over centuries in fairly incremental steps. Romanesque rounded arches gave way to the pointed arcs of Gothic and Islamic design; flying buttresses and stone yielded to internal skeletons and structural fabrics of a lighter kind. Whenever the outer limit of possibility was stretched yet again, a timeless question remained: Will it stand?

In cybernetic architecture, a different situation prevails. As the flywheels of development spin ever faster, there is a dangerous tendency to adopt systems long before they have had time to prove their worth. One untried system is piled upon the next. The overall level of risk soars further up the scale. Supposedly, we can dodge these risks by a more intensive use of computer simulations and highly evolved programs. By and large, this approach works surprisingly well – although the costs of miscalculation are escalating, and everything depends upon the integrity of the source code. One hint of the possible costs came on Wall Street in October 1988, when computerized trading programs nearly triggered a world financial meltdown. We are now invited to accept that this can never happen again, because somebody somewhere has patched up these programs with still further layers of code. Yet there is no such thing as flawless software. An ordinary, off-the-shelf database program, designed by committee to execute just one set of specific tasks, nowadays embodies millions of lines of digits. If printed on paper and stretched end-to-end, they might well span an ocean. Yet, despite the mobilization of some redundant systems and 'smarter' software, all it takes to bring the house down is a *single* misplaced zero or one. Get the picture? The same holds true for semiconductor chips, incidentally. As Intel's chairman, Andy Grove, remarked in 1994, when a flaw was discovered in early versions of his top-of-the-line Pentium chip, there is no such thing as a perfect microprocessor. Yet, today, quicksilver electrical pulses quivering through such fragments of highly refined sand lie at the operative heart of critical systems like jet aircraft, car electronics, advanced weaponry, medical equipment, and many more. The reliability of these systems and our own survival are increasingly enmeshed.*

Consider: many thousands of years ago, as 'backward' civilizations spread through the valleys of the Tigris and Euphrates rivers, they ruled that if a house builder's structure were to collapse he would be liable for prosecution for a capital crime and put to death. As civilizations advanced, they demanded that their architects, doctors, pilots, and lawyers obtain certification for professional competence. Of course, occasionally these professionals still make serious mistakes, but the certification procedures not only minimize the structural risks but also establish a clear framework of responsibility. Today's digitally literate craftsmen of code are

*The design of critical systems is often decisively motivated by short-term profit considerations. The digital avionics now installed routinely on commercial jets were developed not because of any operational benefits they bestow but because they are cheaper than analog avionics and offer savings in weight and fuel costs. What's more, the most thorough designers often work under grueling deadlines: 'We were under tremendous pressure to just ship,' one engineer confided in me. When design takes place in large teams, it is easy to leave double-checking to one's colleagues. The usual problems of coordination are also amplified by the fact that team members may have conflicting priorities.

vastly more influential than the scribes of Babylonia or the priests of ancient Egypt. Yet they do not operate within any of the same strictures and structures that guide all other professions. However crucially our lives depend on their expertise, the success of their hardware and software is measured by commercial viability alone. This may explain why those with a vested interest in mobilizing complex cybernetic technologies consistently underestimate or dismiss the challenges inherent in their safe operation, and neglect to consider their wider sociopolitical effects.

Peter G. Neumann is a computer scientist who advocates a more holistic approach toward technology and who, it is probably fair to say, has done more than almost anyone in the world to raise the level of professional and public awareness regarding the proliferation of computer-related risks. He writes that:

> ... in our lives as in our computer systems, we tend to make unjustified or oversimplified assumptions, often that adverse things will not happen. In life, such assumptions make it possible ... to go on living without inordinate ... paranoia. In computer systems, however, greater concern is often warranted – especially if a system must meet critical requirements under all possible operating conditions to avoid serious repercussions ... [And yet] we still do not appreciate sufficiently the real difficulties that lurk in the development, administration, and use of systems with critical requirements.

Neumann expresses 'optimism that we can do much better in developing and operating computer-related systems with less risk,' but 'pessimism as to how much we can justifiably trust computer systems and people involved in the use of those systems.'[3]

The fact is that we are entrusting too many critical decisions – those that affect our health, our welfare, and the future of our environment as well – to people and processes that work according to rules of their own. The true extent of society's captivation by (and dependence upon) a narrow base of specialists, whose powerful but demonstrably fallible computer algorithms embody specific commercial, technical, and/or political biases, is still only poorly understood, even as the public's dependence upon them verges on complete. The result is a growth of fallibility without corresponding procedures for pinpointing who – or indeed what – is accountable.

After the Second World War was won, the exhilaration of victory quickly gave way to a momentous debate in the most rarefied realms of the scientific establishment. It began, quite literally, in the aftershock of the first atomic blasts, even as the mushroom clouds were still clearing high over Japan. Manhattan Project scientists developed the atom bomb at Los Alamos, in New Mexico, under the direction of the physicist Robert Oppenheimer. Now, for the first time, they were able to set aside the single-minded focus that had helped them meet their technical objective and to ponder instead the wider human meaning of the terrible forces they had unleashed. Many had serious misgivings about the remarkable speed with which an abstract proposition had been translated into reality – one that could blot out any future for mankind. On the night of 2 November 1945, Oppenheimer mounted a podium in Los Alamos to address his colleagues about the 'deep issues ... which involve us as a group of scientists.' He reminded them that development of the atomic bomb had been an organic necessity. He did not seek justification in the fact that the enemy was trying to do the same thing, or protest that free science made for a strong nation more fit and efficient in its prosecution of war. (Such arguments are heard frequently today.) Rather, he said, the bomb had been developed because the essential meaning of Being – of life itself – had actually been threatened by fascist regimes. It was constructed to defend 'a thing worth fighting for and a thing worth living for.'

He continued:

> ... atomic weapons are a peril which affect everyone in the world, and in that sense ... a completely common problem, as common a problem as it was for the Allies to defeat the Nazis. I think that in order to handle [it], there must be a complete sense of community responsibility ... we are not only scientists: we are men, too. We cannot forget our dependence on our fellow men. I mean not only our material dependence, without which no science would be

possible, and without which we could not work; I mean also our deep moral dependence, in that the *value* of science must lie in the world of men . . . [4]

Interestingly, Norbert Wiener, the father of cybernetics, found himself evolving a similar point of view in his work in the emerging worlds of computing and artificial intelligence. 'If we program a machine for winning a war,' he maintained, 'we must [consider] what we mean by winning.' The values of winning 'must be the same values which we hold at heart . . . [and] if we ask for victory and do not know what we mean by it, we shall find the ghost knocking at our door.'[5]

In principle, of course, it is the scientist's high calling to expand the limits of human understanding; in practice, however, this work is too often harnessed to terrible ends. Both Wiener and Oppenheimer felt that the application of technical know-how had to be placed firmly in a social and ethical universe: that knowing *how* could only be meaningful if it was combined with a knowledge of *why*.* More opportunistic Cold Warriors – RAND Corporation men like Edward Teller and John von Neumann – played an altogether more calculated game. They expediently preached a faith in 'science for science's sake.' Pure research should proceed boldly and unrestrained. In the midst of the greatest military buildup in the history of mankind, von Neumann made the flippant

*Both paid dearly for taking these positions. Wiener was ridiculed as an eccentric (if unassailably brilliant) buffoon. Oppenheimer, still chairman of the Atomic Energy Commission's advisory committee in 1949, led the Commission to a unanimous recommendation *against* a superbomb crash development program and in favor of a more open public debate. He was intimidated, reviled in the McCarthy witch hunts, and eventually stripped of his high security clearances.

remark that 'if you say why not bomb [the Soviet Union] tomorrow, I say why not today? If you say today at 5 o'clock, I say why not one o'clock?'[6] As for the widespread concern that 'progress' was spinning out of control, he argued that 'it is precisely technology that enables us to deal with its problems.'[7] Thus, confronting the dilemmas of geopolitical instability, von Neumann and those of similar kidney successfully pressed for an accelerated program of secretive thermonuclear testing and development, combined with almost pathological computer modeling and game theories, in the hope of creating a somehow more rational (and thus more predictable) world. The primary result was a now all-too-familiar military doctrine called 'mutually assured destruction' (or MAD). Confronted with evidence that testing would not only escalate the arms race but also increase the risk of cancer deaths among civilians living in areas adjacent to test sites, von Neumann gave a rather ruthless shrug and replied this was an unfortunate but nevertheless acceptable price to pay for establishing military precedence.* In discussions with his friend the physicist Richard Feynman, he suggested that 'a scientist need not be responsible for the entire world ... social irresponsibility might [indeed] be a reasonable stance.'[8]

How familiar it all sounds today. It is as if, after the Second World War, an ethical seesaw came down with a jarring 'thump' on the side of unrestraint, where it has remained to this day.† Reasoned doubt is transformed into 'Luddite reaction'; those who still persist in steering Wiener's course are lumped together in the same heap with the religious persecutors of Galileo's day: agents of darkness and superstition.

*Ironically, von Neumann's own life was snuffed out in 1957 as the result of a bone cancer brought on by his frequent and enthusiastic visits to nuclear test sites. Having experienced at an impressionable age the considerable fears associated with having to flee from Nazi persecution in 1933, he was left with a lifelong and exceptionally well-developed terror of uncertainty and death. His panicked screams were said to have echoed through the darkened corridors of Walter Reed Hospital during his last days, making a sad coda to his lifetime of extraordinary achievement. Admiral Lewis Strauss, at a dinner in von Neumann's honor, remarked that, 'until the last, he continued to be a member of the [Atomic Energy] Commission and chairman of an important advisory committee to the Defense Department. On one dramatic occasion near the end, there was a meeting at Walter Reed Hospital where, gathered around his bedside and attentive to his last words of advice and wisdom, were the Secretary of Defense and his Deputies, the Secretaries of the Army, Navy and Air Force, and all the military Chiefs of Staff.' (Quoted in Steve J. Heims, *John von Neumann and Norbert Wiener: From Mathematics to the Technologies of Life and Death* (Cambridge, Mass.: The MIT Press, 1980), p. 371.)

†According to one biographer, Steve J. Heims, the flamboyant von Neumann came to be seen as 'a paragon of science and a technologist par excellence.' However, his stance 'raises fundamental issues concerning the scientific community, technology, and our advancing but simultaneously deteriorating civilization.' As Heims describes it, 'On a personal psychological level [von Neumann] was deeply committed to unlimited technological progress, to be achieved with the ↰

greatest possible speed, a commitment related to ... an attitude of cheerful optimism, as if innovation could rejuvenate us, could save us from old age and death, that cyclical, unprogressive aspect of time. This was the irrational foundation that underlay [his] sophistical application of reason. On a sociological level his commitment was to providing high technology for that powerful group sometimes called the military-industrial complex, in effect strengthening the already great power of that group. Uninhibited by ethical considerations in his drive toward innovation ... he chose to be "hard-boiled" concerning the exploitation of nuclear weaponry. By putting himself at the service of the conservative powers that be, he elaborated and helped to advance existing trends in weapons policy and could not help but become a symbol for these very trends.' (Heims, *John von Neumann and Norbert Wiener*, pp. 368–9.)

But – '*pure science*'?

Where do we find all of these dedicated idealists toiling long hours in selfless pursuit of truth? Where are the scientists who, with the devotion of artists, pursue their labors without regard to financial reward? The sacred groves of pure research are few and far between. The overwhelming preponderance of scientific and technological invention is now in fact funded and harnessed by industry. Thinkers like Euclid, Galileo, Newton, and Einstein worked largely in isolation; but ever since Thomas Edison systematized the methods of invention at his now-famous industrial research laboratory at Menlo Park, New Jersey, which opened in 1876, technologists have tended to work in large, coordinated teams. Now they are mainly professionals with a vested financial interest in achieving their programed goals, and the cybernetic systems they generate are leading-edge weapons in a modern arms race for money and markets. As US president Bill Clinton remarked at the start of his first term of office, 'Civilian industry, not military, is the driving force behind advanced technology today ... Only by strengthening our civilian technology base can we solve the twin problems of national security and economic competitiveness ... Leadership in developing and commercializing new technologies is critical to regaining industrial leadership, creating high-wage jobs, and ensuring our long-term prosperity.'[9]

The echoes of history, particularly that of the Cold War, suggest that it is time for technologists to engage in a more meaningful and responsible dialog with the societies that must absorb the consequences of their work. Norbert Wiener believed that 'the independent scientist who is worth the slightest consideration [has] a vocation which demands the possibility of supreme self-sacrifice.'[10] Indeed, after a brilliant start, not least in the US Army's ballistics research and testing grounds at Aberdeen, Maryland, where he made great strides in improving the speed of missile and bullet trajectory calculations, Wiener eventually renounced all work on defense-related contracts at great personal cost. This decision launched him on a writing career that enabled him to make ends meet.

Similarly, Joseph Rotblat, who won the Nobel Peace Prize in October 1995, worked on the Manhattan Project at Los Alamos. He resigned in 1944 when he was tipped off by the project's director that their primary strategic purpose had been secretly reformulated: their now more powerful bomb would be used to threaten Russia in the postwar world. Rotblat co-founded the Association of Atomic Scientists, signed the famous anti-war manifesto drafted by Bertrand Russell with Albert Einstein's support in 1955, and was one of the early movers in the Committee for Nuclear Disarmament. He is persuaded that, although the consequences of a given innovation can never be fully foreseen, technologists are certainly in a better position than most to distinguish and thus act responsibly on the most likely trends of development that lie ahead. Rotblat believes 'everybody has a responsibility for what he is doing. If the work is threatening society in a profound way, how can one say, "That's not my business, I'm just doing my job"? Unfortunately,' he adds, 'most still take this view ... '[11]

In a world supposedly being built on technologies to aid communication, the mechanics of *how* are being disassociated from the ethics of *why*. Under these circumstances, what is the human meaning of such work? As we scramble to adopt manifestly fallible systems in permutations of ever-increasing complexity, scale, and scope, we should also continually ask whether it is still possible to ensure accountability under the new regime.

The cybernetic system is sometimes described as an invisible informational sea whose waves are supposed to reflect — at every point — those of the natural world. But it is more akin to an artificial membrane. Cybernetic systems are being interposed between individuals, and between the whole of humankind and its experience of the natural world. The permeability of this artificial membrane – which is to say its ability as a system to reflect and naturally interact with outside (or 'exogenous') influences – is a function of how it is designed. The cybernetically wired airline is a good example – a microcosm that reflects the broader trendlines affecting both the economy and the polity at large. With the jet-liner, it has now become possible for digital flight management systems (FMS) to control every element from takeoff to landing. The pilot, who is now merely required to 'punch out' any buttons that the computer console lights up for his attention, is meanwhile disconnected from the physical air currents and forces that sustain the craft. As one systems specialist explains:

> Pilots control a conventional airliner's flight path by means of a 'control column.' This is connected to a series of pulleys and cables, leading to hydraulic actuators, which move the flight surfaces. 'Fly by wire' eliminates the cables, and replaces them with *electrical* wiring: it offers weight savings, reduced complexity of hardware, the potential for the use of new interfaces, and, even, modification of fundamental flight control laws – *all made possible by the necessity of having a computer filter most command inputs.*[12]

The pilot's function is now more akin to that of an information manager. Flight data such as airspeed and fuel levels – which were once visible on a diversity of redundant dials and gauges – are now screened as electronic 'pages,' or snapshots, on a single display. This means that it has become more difficult to monitor numerous changes in a continuous timeline: that is, to watch the indicator needles move up or down, each independent of the next, to compare these indicators to the ambient circumstances, and to form an overall synthesis about what this might mean. Instead, by presenting the pilot with discrete fragments of digitized data, all delivered by the same computer via its systems display, reliance on the *automatic* monitoring and programed interpretation is increased, even as the pilot's direct contact with the physical flight elements is stripped away. 'Research suggests that in an emergency situation, pilots of conventional aircraft have an edge over pilots of a modern airplane with the equivalent mission profile, because older [aircraft] require pilots to participate in the feedback loop, thus improving awareness.'[13]

In times of complexity and information overload, whether in passenger airliners or in the metaphorical vessel of human society as a whole, an ability to accurately synthesize and directly act is paramount in the struggle to survive. Yet, ironically, as the world increasingly relies on automatic systems and the specialists who manage them, its overall capacity for effective synthesis, and far-sighted action, seems radically to decrease. Disconnection from the 'control loop' seems to coincide with a trend toward un-managed acceleration, which is especially pronounced in modern combat. A contemporary tank commander who peers through his virtual-reality goggles has no sense of whether the target on his screen is simulated or real. This naturally dilutes his personal engagement when the red trigger button is pressed. How is this likely to affect our survivability in the long run?

Long ago, Zen calligraphy masters recognized metaphorical parallels between their manner of writing and one's wider passage through all of life. A brush is inked, poised, and then applied irrevocably to rice paper in a flowing arc. The resulting character expresses a unique convergence of unrepeatable energies. Hope-fully, there is a perfect harmony between freedom and mastery. Digital simulations, in contrast, disconnect us from this direct commitment to and awareness of the consequences of our acts. Instead, they encourage us to believe in the endless 'repairability' of mistakes. (Indeed, as one digital artist told this author, 'there *are*, in fact, no mistakes.') At the very least, the 'shoot first, think later' mentality rewards reflex rather than reflection: as Wiener said, 'the very speed of operation of modern digital machines stands in the way of our ability to perceive and think through the indications of danger.'[14]

Forthcoming weapons will be even more complex and auto-mated than those of today. They are being designed on the age-old premise that, to engage one's adversaries successfully, speed, strength, and agility should be optimally mixed. Response time should be as close to instantaneous as possible. The control intervals of the computer chip – which cycles in millionths of seconds – already far outpace our own neurological capacity for conscious response. To work at all, therefore, the modern weapons system has to be directed by programed intelligence and coded routines. These systems radically extend our reach and multiply our capacity for devastation. However, they also deny us a large measure of control. They act as proxies, making choices on our behalf despite the fact that they have no conscious will and are unaccountable for the results. What's more, such individual systems – which now guide not just warfare but business as well – are being linked to others in vast distributed networks. The result is that *responsibility* is subtly shifting from analog into digital space – out of a human matrix and into a diffuse web in which its meaning is changed if not altogether eliminated.

For example, the US Aegis and Phalanx missile systems are programed so that they can protect themselves robotically. They can recognize and fire upon incoming targets, making life or death decisions with entirely automated 'judgment' calls. Current fighter aircraft have been deliberately rendered so unstable – to enhance maneuverability – that they simply can't fly without their silicon brains. All in all, battle is coming to seem like 'a chess game played by hysterics on speed,' as one systems designer has memorably said. Battle, once joined, largely proceeds without human agency. Individual commanders have far less scope for individual maneuver in the heat of actual exchange: hence the many unfortunate deaths and casualties associated with 'friendly fire.' (Arguably, such disasters are no longer accidental, in the strictest sense of the term: the machines are simply following their orders.) The human neurological system is simply incapable of reacting in the thousandths of seconds at which computerized systems decide and act. Thus a radical (and sometimes convenient) gulf is opening out between our technical capabilities and our means for ensuring accountability for the resulting events. The commander, the businessman, and the politician, confronted with some unpalatable trend, can now throw up their hands and say, 'It's out of control.' This, then, is one of the great towering ironies of the supposed digital 'revolution': by extending and amplifying human capabilities while at the same time dissolving associated networks of decisionmaking and mutual accountability, its primary effect is to advance a suicidal status quo. It can only be brought under control if its programed goals are redefined.

Oddly enough, the myth still persists that machines are only as good or bad as the people who use them. Karlheinz Stockhausen, the German avant-garde composer, has worked with computer synthesizers for years and he describes his 'instrument' as a mere tool with 'no will at the other end.'[15] However, like many artists, writers, and pundits, Stockhausen works in a self-contained sphere, effectively decoupled from the vast interconnected systems of a cybernetically wired world that are dangerously liable to 'purposeful' life (and spectacular fault). Humankind has indeed reached a state predicted by Norbert Wiener three decades ago. He warned that 'by the very slowness of our human actions, our effective control of our machines may be nullified.' The consequent danger is that 'by the time we are able to react to information conveyed by our senses and stop the car we are driving, it may already have run head on into a wall.'[16]

Ancient cultures recorded their knowledge on clay tablets, papyrus, and stone. The few fragments of this information that have come down to us today provide an insight into our historical origins and also satisfy an innate fascination with how others, before us, actually lived. It took a great deal of ingenuity to decipher the Dead Sea Scrolls and the hieroglyphs of pharaonic Egypt, but the existence of physical artifacts and references meant that it could eventually be done. There is a danger that the symbolic output of our own civilization will in the future seem like nothing more than white noise.

By the end of this decade, the overwhelming preponderance of information, now printed, will probably be created in exclusively digital form. Fortunately, this will help libraries cope with the soaring costs of physical storage. However, Charles Dollar, a director at the National Archives in Washington DC, is still concerned. He points out that the software programs controlling data compression, elaborate cross-referencing and hypertext applications are mainly proprietary.* What's more, they are evolving and being replaced all the time. So, too, are the data-reading devices and magnetic storage media, which ironically enough seem to wear out much faster than pulp. At this point, he concludes, we have no guarantee that future generations will actually be able to retrieve and make sense of the records and experience being created and stored in bitstream form today.[17] Moreover, in the absence of analog originals, there is no way of scientifically verifying the integrity or authenticity of digitally defined information.

*Note on hypertext: reading the word 'France', your mind might spontaneously generate the phrase 'Paris in the spring', and conjure a series of associated images, instead of continuing to read the words as they appear before you on the page. Your mind has made a lateral leap. Such associative leaps can be written into electronic text through the use of 'links.' Thus a programmer can assign a series of links to the word 'France' so that clicking on that word with a mouse produces a series of predefined associations. The difference between analog and digital links, in this case, is clear: the one is spontaneous and personal, the other reflects how the system is designed. See the footnote on CGI scripts on page 220.

The integrity of the historical record is crucially enmeshed with the systems by which it is coded and conveyed. Apple Computer, which has carved out a strong niche in the education market, is enthusiastic about the proposition that, once enough information has been digitized, software can replace physical classrooms. This will free students to go wherever their minds may lead (or be led). John Sculley once remarked that digitization will bring 'into question whether the institutional experience of education should be bound by walls of brick or whether it should be something that can go wherever the students go.'[18] Parents will be assigned 'vouchers' to buy the educational 'services' they believe their children require, chosen according to their own individual and ideological predilections and consumed in the home. Alan C. Kay, an Apple Fellow, suggests that, by the year 2000, 'powerful, intimate computers will become as ubiquitous as television and will be connected to interlinked networks that span the globe more comprehensively than today.' If used properly, he says, scholastic computers can be 'powerful amplifiers, extending the reach and depth of the learners.' However, he goes on to admit, 'network computer media will *initially* substitute convenience for verisimilitude, and quantity and speed for exposition and thought-

fulness.'[19]★ Note this recurrent
theme, like the percussive thump
that drives a disco beat: it is
agreed that *initially* there will be
some problems – those caused by
our new tools – but *eventually* everything will be put technically straight.
In the meantime, older and more

★Of course, the visual immediacy of digital
teaching tools will be a poor substitute for
direct and tangible experience. But, for those
who might never have the chance to see
Botticelli's *Primavera* in real life, access to a
video-disc simulation of the painting is a positive thing, of course, even if that version will
never capture the elusive mystery of the
original as closely as a good traditional print.

reflective media will be marginalized, and there is a real danger
that the classroom experience may be radically commoditized.
Similarly, a general overreliance on digital artifacts may stifle imagination, individual initiative, and creative growth.

The digitization of our archives and of the modes of education is
thus symptomatic of a much wider concern: how do we select our
programers? Authenticate informational integrity? Verify that we
are pursuing the intended goals? (For instance, if all aspects of
education become subject to private consumer choice, how can the
binding values of a society survive intact?) As head of the US
Library of Congress's interactive media project, Jacqueline Hess
posed the most pertinent question: 'Who *moderates* all this information? Where's the commonality and overarching quality control
that gives kids a sense of historical context when they're exposed to
Adolf Hitler's *Mein Kampf*?'[20]

Or, more to the point, how can we be sure whether a Hitler has
been properly voted in and did not actually steal the election – that
is, that the results we are presented with are actually valid? A scene
springs to mind: a yellowed photograph clipped from a newspaper
long ago. Two grizzled men are shown descending a snowy hillside
with a heavy ballot box swinging between their legs. Each has a
rifle slung over his arm. They have come from a high mountain
village in Turkey which is split by rival political clans. These men
have been chosen to deliver the town's ballots to the valley: they
are the town's trusted intermediaries. It is their task personally to
validate that the data are delivered intact – and to confirm that the
results of the election are correct.

Kevin Kelly, of *Wired* magazine, believes that such roles can be performed automatically by complex cybernetic machines. He celebrates the metaphysical advance of automatic control in human history. First came the control of steam and other forms of energy. Next came the control of materials: smaller mechanisms shot through with computer chips can do the same work as much larger 'dumb' ones (like ourselves) did before. Then:

> The third regime of the control revolution ... is the control of information itself. The miles of circuits and information looping from place to place that administers the control of energy and matter has incidentally flooded our environment with messages, bits and bytes. This unmanaged data tide is at toxic levels. We generate more information than we can control. The promise of more information has come true. But more information is like the raw explosion of steam – utterly useless unless harnessed ... Genetic engineering (information which controls DNA information) and tools for electronic libraries (information which manages book information) foreshadow the subjugation of information. *The impact of information domestication will initially be felt in industry and business*, just as energy and material control did, and then seep later to the realm of the individual ... Investing machines with the ability to adapt on their own, to evolve in their own direction, and grow without human oversight is the next great advance in technology.[21]

Should civil societies be prepared to trust such programed agents in applications that are critical to the health of society, the vitality of economic life, and the continuity of democracy itself? For instance, do we want electronic polling stations that deliver the outcome of polls without any means for public scrutiny? Of course not. Yet in manifold (if sometimes more subtle) ways digital technology is presenting this underlying question: To whom (or what) are we passing the role of 'trusted agent'? Today we almost automatically assume that the systems that surround us are essentially benign, and that they will work, as advertised, on our behalf. Microsoft's Bill Gates paints an almost utopian vision of ubiquitous computing:

> Your wallet PC will be able to keep audio, time, location, and eventually even video records of everything that happens to you. You

will be able to record every word you say and every word said to you, as well as body temperature, blood pressure, barometric pressure, and a variety of other data about you and your surroundings. It will be able to track your interactions with the [information] highway – all the commands you issue, the messages you send, and the people you call or who call you. The resulting diary will be the ultimate ... autobiography ... [22]

However, Mr Gates's discussion sidesteps a related and more powerful concern. What if such systems become obligatory to efficient economic life? Do we wish to have such diaries floating around? How about that little thing called privacy? Mr Gates points out that 'everyone is willing to accept some restrictions in exchange for a sense of security.'[23] Now, this is probably a perfectly well-meaning thought; the Devil, as ever, lies in the details. It sounds suspiciously like that historically familiar Faustian bargain in which a people, fearful of instability and/or overrapid change, agree to place their faith in systems that promise peace of mind and insulate them from the root causes of their concern. Such trade-offs can be accepted only in a spirit of unrelenting vigilance, for they imply personal and moral surrender to technical paternalism, and constitute the thin end of a wedge that can all too easily be used to prise open a door to authoritarian rule.

The general public and its advocacy groups should be asking privacy-related questions with far more urgency and persistence. A fully wired world – a landscape we cannot efficiently negotiate without digital money and a complete array of chip-based tools – is also potentially a world in which our every act is recorded and taken into account. Who (or what) will actually prevent forthcoming wall-mounted flat-panel displays – also called 'interactive' TV – from becoming the ubiquitous telescreens of George Orwell's *1984*, with all the attendant commercial (rather than political) oppressions involved? Is privacy indeed *possible* if electronics are embedded in everything from blue suede shoes to brass doorknobs – and when swarms of intelligent software 'agents' can traverse the Web at personal or corporate behest? Will insurance companies ferret out the results of our medical tele-consultations and cancel our policies if we prove ill?

These are not abstract concerns: they are already very real and immediate.

Netscape, the popular Web browser, includes a feature which enables companies to find out who their customers are, and how they behave when logged onto a given Web site. This information is organized in nuggets, called Magic Cookies, invisibly installed in the customer's own computer hard disc. Their form is fully readable by the remote company alone. The ominous potential of Cookie-like tools can easily be masked in the impenetrably technical language in which their specifications are described. For instance, in the Netscape case, Cookies are described as 'a general mechanism which server side connections (such as CGI scripts) can use to both *store and retrieve information on the client side of the connection.* The addition of a simple, persistent, client-side state significantly extends the capabilities of Web-based client/server applications.'[24]

Is that perfectly clear?* Meanwhile, according to *Information Week* magazine, 'Microsoft officials confirm that beta versions of Windows 95 include a small viral routine called Registration Wizard. It interrogates every system on a network gathering intelligence on what software is being run on which machine. It then creates a complete listing of both Microsoft's and its competitors' products by machine, which it reports to Microsoft when customers sign up for Microsoft's Network Services (MSN), due for launch later this year.'[25] As one contributor to the comp.risks newsgroup on the Internet notes, using the free demonstration of MSN 'transmits your entire directory structure in the background. This means they have a list of every directory (and, potentially, every file) on your machine.'[26] While technical details change, the dangers remain.[27]

Bill Gates believes that 'if ubiquitous cameras tied to the information highway should prove to reduce serious crime dramatically in test communities, a real debate would begin over whether

*'CGI' stands for 'Common Gateway Interface'. It is the Internet standard that conditions how a Web client (such as a 'user') can interact with Web servers (which are usually, though not always, run by governments and commercial enterprises). For example, such a mechanism, as implemented on a software-company server, might shape the possibilities of interaction in such a way as to steer a user away from testing an older or less lucrative product and toward a newer and more profitable one.

people fear surveillance more or less than they fear crime ... what today seems like digital Big Brother might one day become the norm if the alternative is being left to the mercy of terrorists and criminals. I am not advocating either position – technology will enable society to make a political decision.'[28] (Again, note this reliance on the agency of intermediating machines to solve problems that are human at root.) Mark Weiser, formerly head of Xerox's Palo Alto Research Center in California, was one of the leading exponents of ubiquitous computing. Now an independent entrepreneur, he still admits to a nagging concern that 'hundreds of computers in every room, all capable of sensing people near them and linked to high-speed networks, have the potential to make totalitarianism up till now seem like sheerest anarchy.' [29]

Light filters through a photographic lens. It settles upon photographic film, rearranges its particles, and forms a physical imprint that we call a negative. For the last hundred years, this very physicality has made photos uniquely credible. 'Seeing' – as we used to say – 'is believing.'

Of course, analog photos could be manipulated as well. Recall the fate of Leon Trotsky. Second only to Lenin in the Soviet Union's revolutionary hierarchy, he was forced into exile in Mexico in 1929. One fine morning some months later – surrounded by frangipani and brightly plumed birds – he opened a copy of *Pravda* that had just arrived from Moscow in the mail. He learned that he had been further demoted in the ongoing power struggle sweeping through the Politburo, and that his leading role in the 1917 uprising was being brushed off as that of a 'counter-revolutionary plotter.' And there was a telling photo that accompanied the text. Carefully doctored by the Kremlin's cut-and-paste specialists, it showed a suggestively blank space where Trotsky had once stood at the very core of the coup leadership. With a bit of time, this revision acquired the force of legitimate historical 'fact.' Then Trotsky was assassinated.

What Moscow's henchmen accomplished with that retouched photograph was a clumsy business involving brushes, razors, and glue. There was inevitably something 'off' about the results: anyone could scent the fake.* Now that photographs can be taken and processed digitally, the situation is changed. With trivial ease, a photo, or part of a photo, can be manipulated in an infinite variety of undetectable ways. The same is true for video 'footage.' An Associated Press article on the promotional placement of commercial products in films cites an example in the blockbuster *Demolition Man*. In the North American version of the movie, one character (Sandra Bullock) informs another (played by Sylvester Stallone) that Taco Bell, the US fast-food franchise chain, actually survived a cycle of low-level wars and disasters into the twenty-first century (in which the movie takes place). In versions of the film seen abroad, the Pizza Hut franchise appears instead. Interestingly, both chains are owned by the same parent company, but, while there are roughly 4,600 Taco Bell franchises in North America and only a handful abroad, there are 3,300 Pizza Hut franchises overseas. According to the AP, special-effects experts digitally removed the Taco Bell logo from the American film, added a new restaurant, and inserted re-recorded versions of Bullock's dialog for the international release.[30]

*It would be foolish to suggest that photographs have never been set up and subsequently passed off as authentic events: witness the argument over whether Robert Capa's photo of a soldier being shot in the Spanish Civil War was staged. But this, as well as more skillful examples of Kremlin-style retouching and exploratory artistic collages by the likes of Man Ray, is the exception that proves the rule. For years the authority commanded by photographs was rooted in the fact that they captured physical events.

In another gray area between the virtual and the real, a new system for processing live TV images was mobilized during the 1996 Atlanta Olympics. Millions around the world watched the same sporting events unfold. But, while viewers in Japan saw a computer-generated logo that read 'Matsushita' in the space directly beside the scoreboard, Americans read 'Microsoft' in the very same place. This allowed broadcasters to maximize their advertising revenue by selectively adjusting the background hoardings to specific national or demographic taste. Only part of what viewers saw physically took place.

The power seamlessly to mix the literal and the virtual, the physical and the metaphorical, builds on processing technologies that have been used by movie producers and advertisers for many years. Audiences understand that films and ads are being altered to achieve some explicitly desired effect. Fred Ritchin, who was director of photography at the *New York Times*, described his first encounter with a powerful Scitex 'Visionary' image-processor, in which he and a few colleagues playfully transformed Manhattan's skyline into a fantastic collage:

> We added the Eiffel Tower, we added the TransAmerica Pyramid from San Francisco, we turned the top of the Citicorp Building on the right [and] moved ... the Empire State Building uptown a few blocks and made it taller. The technology is amazing ... you have this impulse to make it bigger, make it smaller, make it pink, because it is so easy. This is what some people have called *The God Complex*. You see yourself rearranging skyscrapers ... you can do whatever you want to do. Just as there is a tendency to manage text, there is a tendency to manage imagery.[31]

What happens when information management is combined with these powerful new tools and spills over (for example) into public affairs? When one candidate takes voice and video clips of his or her opponent, and remixes both out of context to achieve some discrediting effect? What rules apply when the mixture of fact and fantasy ceases to be overt – as it is with a film which we go to see specifically for its special effects – and is deliberately disguised instead? As we skate over more simulations, interfaces, and visual media that obscure the programed complexities beneath, by what means do we ensure their ultimate accuracy? Food companies have to meet truth-in-advertising and product-labeling requirements today. Should they also be obliged clearly to caption any images in promotional 'advertorial' programing that have somehow been digitally fixed? Reality – translated into binary streams – is an infinitely movable feast. Power – unchecked – is absolute.

One of the great strengths of the photograph was that, although it was the product of a single point of view, and was achieved with the wide variety of technical means at the command of a photographer, it nevertheless recorded authentic events. Here was an old analog technology that could be employed by the weak to redress their disadvantage against the strong. For well over a century, it has therefore served not only as a potent tool of propaganda but also as a means of witnessing and of confronting brute power with the consequence of its acts. The horrors of the Nazi death camps; the Tiananmen massacre and the Chinese government's feeble attempt at denial; the Rodney King police-brutality case in Los Angeles: photos and videos depicted realities that directly challenged the comfortable and prevailing 'truths.' This witnessing role – the natural *datum* or physical reference point – is being scrapped. When the authenticity of an image cannot be verified, its empowering qualities erode. The most persuasive signal managers will arrive to fill the empty space.

For a short interval in human history, when confronted with a picture that challenged one's world-view, a person felt the compulsion to rethink his or her position. Now, Fred Ritchin believes, 'the so-called "average" reader may no longer accept photographs the way he or she used to.'[32] If an image does not agree, readers may simply say, 'Well, the photograph must be wrong, it must have been manipulated by those computers of yours.' They may place their faith in the simulation of their choice – one free from the disciplines of natural fact. Indeed, as the advanced technologies of digital imaging are translated into affordable head-mounted virtual-reality displays – the video Walkman – image-saturated Western consumers will soon find themselves confronting an unprecedented new sensory phenomenon. Like airborne trapeze artists, we will find ourselves grasping at equally plausible but mutually exclusive alternative versions of the very same 'event.'

From the dawn of history to the start of our own digital day, we have been inventors of and workers-upon of things: *Homo faber* – the magician of the hand and mind. With time, our creations have grown more complex. Our artifacts are woven into the tapestry of daily experience – we use them without a second's thought. They become – in effect – invisible. Alfred North Whitehead, in his *Introduction to Mathematics*, wrote that 'Civilization advances by extending the number of important operations which we can perform without thinking about them.'[33] But life in a world of scarcity will demand a new approach. Humankind will need to grow *more* aware — to exert more hands-on care — than ever before. Yet an increasing number of the implements with which we furnish our lives are impervious to the care of hand or mind. They present either an impenetrable or a transient face. When a digital watch, an electronic appliance, a stereo component, or a computer decides to break down, it's often far more sensible to buy a new one than to try to have it repaired. Cars are jammed with electronic control systems, but the notion of pulling out a few tools and fixing a failed anti-lock brake unit yourself is now a wistful dream. Kiss the Zen of motorcycle maintenance goodbye. Make friends with a highly trained technician. Check your digital bank balance. And pray.

We use our advanced systems with a level of naive trust that is indeed unique to an advanced civilization. It is only when an aircraft crashes, when the stock exchange falters, or when the Internet takes a dive that our boundless faith is undermined. These events confront us with the sheer depth of our technical dependence. The author John Hockenberry mused on this addiction at the time of Intel's Pentium debacle:

> The unsullied clarity of being sure beyond doubt that the solutions to my physics problems were correct is a feeling from the days when I could still double-check my calculator; it has been replaced by doubt and powerlessness. The capacity to believe in absolute truth is a priceless gift we can value now that we have worked so hard to lose it. We stare at the laptop, like Adam and Eve, trying to find a way back to the garden and hoping we can still run Windows if we get there. Perhaps the error is not with the Pentium chip.[34]

In fact the error lies with those who are like sailors who go to sea equipped with all the latest Global Positioning Systems and computer map displays (more often than not based on charts drawn up by the British Admiralty and US whaling crews in the nineteenth century) yet neglect to bring a backup compass and sextant (neither of which has chips to break down or batteries that inevitably run dry). It is terribly easy to grow comfortable and forget that the marine environment is unforgiving of delicate things. The approach from Europe to the West Indies and the New World is treacherous with reefs. Likewise, in crossing the Digital Ocean in pursuit of new frontiers, there is grave danger in scrapping redundant backup systems, on which all our ventures now depend, merely because of their implied costs and 'inefficiency.' If we jettison direct contact with our tools, if adequate understanding of these systems fades away, a crackup on the reefs is only a matter of time.

In 1986 I traveled through the Norwegian Arctic on the M/S *Polarys*, a coastal mail packet built just after the Second World War. She is just one of a coordinated fleet of steamers that sail up and down the rough, fjord-pocked coast – part of the infrastructure that helps weave Norway's collective psyche into a coherent whole. Outside the windows, great peaks of rock and ice floes passed with spectacular and frigid indifference. Inside the steamer's staterooms and salons there were substantial fixtures of brass that glowed with years of care. The hardwood panels had a warmth – almost an animation – that only endless rubbings with beeswax polish can impart. Sailing aboard this old diesel vessel was like holding a handwritten dispatch from an old friend in the palm of your hand. It filled you with a sense of *prana* – a word used in India for that mysterious breath of spirit that flows through the life of all things.* It is a breath that always seems strangely faint in items untouched by Nature's hand. This ship – built to endure, and cared for well – had *prana*. She was a far cry from the countless products we cannot repair – and from the thoughts and experiences that become like Styrofoam clamshells

*A similar concept animated the writings of William Morris in the Arts and Crafts movement, which tried to pose an alternative to industrial mass production a century ago. Today, more than ever before, the touch of human craft is becoming reserved for the privileged few.

we are expected to dispose of once their purpose is served. When a fusion of economics and technology makes it cheaper to shed rather than to repair or care for the responsibilities we have assumed through creation – indeed, when the very design of our works prevents interaction with them in any meaningful way – then it will be impossible to set things right when the train of cybernetic society veers off its track.

Of course, we can still hope for the best. Who knows? Maybe John Perry Barlow is right when he says that groundless hope – like unconditional love – is the only kind that ultimately counts.[35]

10

Cybertrends

Think what it would be to have a work conceived from outside the self, a work that would let us escape the limited perspective of the individual ego, not only to enter into selves like our own but to give speech to that which has no language ... [perhaps this is] what Lucretius was aiming at when he identified himself with that nature common to each and every thing?

ITALO CALVINO [1]

In Mario Vargas Llosa's 1996 novel, *Death in the Andes*, an internationally respected naturalist is brutally slain by a group of terrorists known as Sendero Luminoso (or Shining Path). This elderly woman makes the mistake of staying behind to work in the jungle after civil order has collapsed and the territory is occupied by armed guerrilla fighters. Having dedicated her life to studying and preserving Peru's natural heritage, she naively believes that she will be kept safe and free from harm. When she is in fact captured by the revolutionaries, she protests that she and they share a common cause. Surely they can see? 'Our concern is nature,' she cries, ' ... we work for all Peruvians.'[2] The angry guerrillas are blind to the idea. Naturalist or no, the woman is considered an embodiment of the cosmopolitan and oppressive ruling class. In other words, she is regarded not as a human being but as a symbol – like a binary zero or one. In the inverted logic of the conflict, this innocent and accomplished woman is condemned to death. A similar fate befalls a pair of visiting French tourists.

In extreme conditions, people find it hard to converse in rational terms. Vargas Llosa's novel describes how many of the world's people are reacting to the tragic *disconnection* between those who most benefit from unrestrained scientific 'rationalism' and those forces of life that have been suppressed in its name. This disconnection has been radically amplified by cybernetic means. Does it come as a real surprise that, lacking dialog, the world finds itself moving down the path of bullets and guns? In the final humiliation, the colonized adopt the barbaric language of their colonizers. This is not just a phenomenon of South America and other 'underdeveloped' parts of the world: it rattles through the large conurbations of Western democracy just as well. Just as pockets of wealthy North coexist with impoverished South, zones of civil order coexist side-by-side with areas of terrible and bewildering violence.[3] It somehow seems, as Italo Calvino once remarked, that 'the more enlightened our houses are, the more their walls ooze ghosts. Dreams of progress and reason are haunted by nightmares.'[4] Michael Heim, a contemporary American technologist and philosopher, is convinced that 'when civilization reaches a certain degree of density, the barbaric tribes return, from within ... A global international village, fed by accelerated competition and driven by information, may be host to unprecedented barbarism.'[5]

Against this background, we are asked to accept that electronic networking will be an unequivocal force for something called 'progress.' Yet it may be precisely this mode of progress – this incapacity to embrace life in the wholeness of its rich relativities – that is responsible for the difficulties we face. Microsoft's Bill Gates cheerfully admits he has a vested interest in optimism about the technologies he creates. He asserts that 'the information highway [will make] all communication easier.'[6] Why should this be the case? The technical capacity to move data hardly facilitates communication, which is essentially a spiritual act: it simply increases the volume and speed of messages under way. When the whole world consists of data, then it becomes a digitally movable feast. Who will move it? Who will *own* it? It has become possible to indulge the ancient acquisitive instinct much faster, over wider distances, and with greater success than ever before – thanks to the amplification of cybernetic techniques. The outcome is a species of untempered 'rationality' run amok.

For instance, Bill Gates is convinced that:

> [the digital revolution] will give us more control over our lives and allow experiences and products to be custom tailored to our interests. Citizens of the information society will enjoy new opportunities for productivity, learning, and entertainment. Countries that move boldly and in contact with each other will enjoy economic rewards. Whole new markets will emerge, and a myriad new opportunities for employment will be created.[7]

Of course, Gates and the team that helped him write his book *The Road Ahead* have acknowledged that 'the benefits of the information society will carry costs.' Moreover, they add with admirable understatement, there are 'equity issues that will have to be addressed.'[8] Nevertheless, 'one of the wonderful things about the information highway is that virtual equity is far easier to achieve than real-world equity'; for example, 'It would take a massive amount of money to give every grammar school in every poor area the same library resources as the schools in Beverly Hills. However, when you put schools on-line they all get the same access to information ... '[9] The net effect of an information economy 'will be a wealthier world, which should be stabilizing ... [and] the gap between the have and have-not nations will diminish.'[10] By early in the next millennium, predicts Nicholas Negroponte, the MIT-based wizard of all things digital, your 'right and left cufflinks or earrings may communicate with each other by low orbiting satellites and have more computer power than your present PC.'[11] Needless to say, the whole of the world's population will be decked out in golden ear bobs and bestudded French cuffs: essential fashion accessories that will have critical messages that they need to convey instantly. In this digital world, Negroponte asserts, 'physical space will be irrelevant and time will play a different role.'[12]

It might be interesting to hear General Norman Schwartzkopf's derisive snort. After all, advanced weapons are designed to burst *precisely* in crosshairs in time and space. Time and space are decisive if you are scrambling to shift tons of men and physical material to the Gulf, as Schwartzkopf did during the 1991 war, so that the US and its allies could expel Saddam Hussein's Republican Guard from the rich oilfields of Kuwait. In fact the whole Gulf War suggests a somewhat different perspective from that offered by Messrs Gates, Negroponte, and their ilk: namely, that however significant the role of information-processing technologies in the overall economic mix of the decade to come, physical commodities will grow *more* rather than less relevant with time. The world will need an additional 15 million barrels of Middle East oil each day just to slake the additional thirst of the world's new automobiles over the next decade and a half.* Moreover, the general increase in raw-material consumption is expected to rocket along with the developing world's industrial growth.[13] (This will of course accelerate the confirmed buildup of atmospheric carbon dioxide, which between 1959 (when systematic measurement began) and 1994 grew by 14 percent.)

*According to the OECD's *World Energy Outlook – 1995* (Paris: International Energy Agency, 1995), total world demand for oil is expected to rise from 68 million barrels a day (b/d) at present to some 95 million b/d in the year 2010, much of which will be sourced in the Middle East. See also Edward Mortimer, 'Trapped between freedom and despair.' *Financial Times*, 23 March 1995. The 'emerging' world's use of oil climbed from one barrel in six of total global consumption to one in four in the decade to 1996 (Robert Cozine, 'Emerging markets lead energy demand growth,' *Financial Times*, 19 June 1996).

Homo sapiens now consumes more than half the resources needed to sustain all of planetary life.[14] Life-sustaining biological diversity, in both wild and domesticated form, is steadily dwindling.† Forests are being felled at a rate of over one acre per second worldwide. And there's more: deserts expand at a rate of 6 billion hectares per year, while the area covered by tropical forest annually dwindles by 11 billion hectares.[15] In all of the world's important food-production areas, excessive demand for irrigation is radically lowering water tables. In many parts of the world, including the already volatile Middle East, low-level water-based conflicts are already under way.

†Roughly a third of the world's 4,500 *domestic* animal species 'are now at risk of loss' from intensive monocropping, writes the UN Food and Agriculture Organization in its *World Watch List for Domestic Animal Diversity* (Rome: FAO Publications, 1995). According to David Richardson, 'most of the calves in the world may soon be related to just a few dozen superior bull families as breeders select from an ever-reducing gene pool' ('Unnatural selection,' *Financial Times*, 3 January 1996).

It is a disturbing litany. It suggests that we badly need to cultivate a more holistic sense of our place – and that of our technologies – in the context of the natural world on which we depend. Paul Hawken, a California-based author and businessman, observes that:

> if you take a basketball and pretend it is the earth, and then paint it lightly with a spray can, the thin emulsion of pigment coating the surface is ten times thicker, relatively speaking, than the band of life that supports our existence on this planet ... the problem of carrying capacity lies not just with the obvious examples seen on our TV screens, but is worldwide. Those who argue that we need to grow our way out of ecological problems do not acknowledge a profound and troubling contradiction: If the population of China lived as well as the population of Japan or France or the United States, we would endure untold ecological devastation.*

These realities rarely seem to figure when the futurists take to the boards with their vision of a new world: a place where everyone is free from time and space. What they are really talking about has nothing to do with metaphysics. It has to do with the development of new markets and new money-spinning products like in-car digital navigators. And since this market will never take shape unless people continue to drive cars, any serious talk about whether the continued unrestricted use of cars is really such a wise idea must be ruled out of order.

*Paul Hawken, *The Ecology of Commerce* (New York: HarperCollins, 1993), pp. 207–8. To reduce the human impact on the earth, Hawken proposes means of production that radically eliminate waste (the 'waste equals food' principle), a change from a carbon-based economy to one built on hydrogen and sunshine (which would generate many new jobs), and 'systems of feedback and accountability that support and strengthen restorative behaviors, whether they are in resource utilities, green fees on agricultural chemicals, or reliance on local production and distribution' (Ibid., pp. 209–10).

The whole specter of the earth's failing vital signs is about as welcome as the sudden appearance of an exasperated spouse to an unfaithful partner indulging in a bacchanalian fling, or the arrival of a winged messenger with troubling news. There are the marine biologists at the UN Food and Agriculture Organization (FAO) who report that *all* of the world's major oceanic fisheries are being fished at or beyond capacity. There is the fact that food production increases are no longer keeping pace with the rate of population growth. (Indeed, in 1996 world grain stocks fell to their lowest levels in two decades and the resulting price rises devastated those least able to pay.*) Barring an extensive plague or some unthinkably terrible natural catastrophe, and lacking more widespread family planning, the world's population is on track to swell from the present 5.5 billion to almost 9 billion in one generation. Lester R. Brown, president of the Washington DC-based Worldwatch Institute, is widely respected for his painstaking attention to factual detail in the sometimes alarmist environmental debate. In his balanced assessment, issues of food security are shortly set to replace those of military security: 'The deteriorating relationship between us and our natural support systems, and the economic effects of that changing relationship, will become a consuming concern of [national] governments in the decades ahead.'[16]

All of these very real and increasingly urgent challenges seem so pleasantly distant to the digital video-game fantasists of our time. Their horizon is untroubled by the fact that, over the next decade, a substantial proportion of the world's population will find its lifestyle radically displaced in that exodus called 'restructuring,' whereby farmers are driven by economic necessity from the hinterland to the metropolis in search of dwindling opportunities. If their continuities erode faster than tangible new values arise to take their place, and if the meanings derived from physical labor are not superseded by equally resonant alternatives, then the result will be stress and cognitive dissonance on a terrible scale. It is all well and good, for those of us who happen to enjoy life in an advanced democracy, to profess exhilaration at this 'liminal moment' in history when all the familiar institutional relativities are

*In 1995, the price of basic cereals – wheat, rice and corn (maize) – increased between 30 and 50 percent and thus imposed an extra $3 billion on the already crippling food bill of the 'developing' world, where almost 1 billion people are already chronically malnourished (Deborah Hargreaves, 'FAO warns of crisis in world food supplies,' *Financial Times*, 2 February 1996). During those same two decades, the international refugee population rose 88 percent, to 26 million. See Lester R. Brown, Nicholas Lenssen, Hal Kane, abstract of the Washington DC-based Worldwatch Institute's *Vital Signs: 1994–95* (Washington DC: Earthscan, 1995). The abstract, incidentally, appeared on the Web at http://www.worldwatch.org/pubs/vs/vs95/index.html.

starting to change. But, outside of our virtual cocoon, is it really a *creative* transition or *productive* stress for the bulk of the world's souls? What happens when you try to load and run a Pentium-based application program onto a system built around the old Intel 8080 chipset? The computer will merely slow or crash – and its data-packed hard drive can always be replaced – but human 'systems,' simply unable to process an overload of contradictory signals, may burst into a cycle of tragic violence.*

A chorus of protest from stage right. Really – such Malthusian pessimism! Information-processing technologies will surely surprise us. They will produce more growth and avert this unspeakable fate. However, neither the tools nor the growth they may generate address the unsustainability of the expansive agenda nor the basic *distributional* dilemmas

*Researchers at Oxfam warn that 'just as crime and social breakdown in the industrial world will not respect the boundaries of affluent middle-class suburbs, so the forces unleashed by conflict and global poverty will not respect national borders, however well-defended they may be' (*Oxfam Poverty Report* (London: Oxfam, 1995), quoted in Michael Holman, 'Global poverty threat to stability, warns Oxfam,' *Financial Times*, 21 June 1995).

that grow ever more acute. Although, as David Korten notes, material expansion is still perceived as 'the key to ending poverty, stabilizing population, protecting the environment, and achieving social harmony,'[17] it has actually failed on every count to date. In fact, it 'is accelerating the breakdown of the ecosystem's regenerative

capacities and the social fabric that sustains human community; at the same time, it is intensifying the competition for resources between rich and poor – a competition that the poor invariably lose.'[18]†

John Gray is inclined to agree. A respected political theorist, formerly associated with the UK's New Right, he has since turned his back on the traditional left–right political dichotomy in search of a more meaningful and genuinely conserv-

†Despite a fivefold increase in world economic output since 1950, writes Korten, 'the number of people living in absolute poverty has kept pace with population growth: both have doubled. The ratio of the share of the world's income that went to the richest 20 percent and that which went to the bottom 20 percent has doubled. And indicators of social and environmental disintegration have risen sharply nearly everywhere. Although economic growth did not necessarily create these problems, it certainly has not solved them.' (David Korten, *When Corporations Rule the World* (London: Earthscan, 1995), p. 39.)

ative agenda. He remarks that 'the industrialization of the world on the model of its richest nations is a dystopian fantasy ... perhaps the most vulgar ideal ever put before suffering humankind. The myth of open-ended progress is not an ennobling myth, and it should form no part of conservative philosophy ... ' What's more, he adds, those who rest their faith in the combination of free markets and technical tools are making a dangerous mistake:

> The growth of technology itself depends on *human institutions* that are always unstable and often desperately fragile; it is disrupted, or retarded, when human institutions break down. For this reason, the growth of technology cannot be guaranteed, and a technical fix for the problems of humankind, even supposing it to be possible, will always be beyond its reach.[19]

Even the World Bank tacitly concedes this point in its draft 1997 World Development Report. It writes that 'an effective ... state is needed to create the institutional infrastructure for markets to flourish.' It believes 'a significant role for the state in the 21st century will be in providing a framework under which the trade-off between market-based growth and escalating environmental problems is resolved.'[20]

The fashion today is to submit to a 'destiny' imposed by machines. As the talented Kevin Kelly suggests, we have already grown far too dependent upon distributed electronic systems to control our own fate. Our creations are taking on an almost biological complexity and a 'will' of their own: man and system exist in a vast and inescapable state of symbiosis. Moreover, machines are on the verge of gaining the upper hand. The only sensible response will be to free systems to evolve of their own accord. Kelly cites the Chinese mystic Lao-tzu, who wrote 2,600 years ago that 'intelligent control exerts influence without appearing to do so.'[21]

In Kelly's considered view:

> ... pervasively networked computers will be the main shaper of humans in the future. It's not just individual books we are leaving behind ... global opinion polling in real-time 24 hours a day, seven days a week, ubiquitous telephones, asynchronous e-mail, 500 TV

channels, video-on-demand: all these add up to the matrix for a glorious network culture, a remarkable *hivelike* being.

The tiny bees in my hive are more or less unaware of their colony. By definition their collective hive mind must transcend their small bee minds. As we wire ourselves into a hivish network, many things will emerge that we, as *mere neurons in the network*, don't expect, don't understand, can't control, or don't even perceive.[22]

One scientist, Marcus Viermenhouk, disagrees. 'Bees are pretty stupid little shits,' he laughs; moreover, 'you can only say so much by shaking your ass.'[23] Human beings, in contrast, have the capacity to behave with civilized intent. They have the potential to devise a more enlightened path ahead. Whether they find the presence of mind to do so in a hive of twenty-four-hour virtual distractions remains to be seen. Bill Gates concedes that emerging networks of virtual escape may indeed grow a tad too *diverting* for some. But never fear: 'If you were to find yourself escaping into these attractive worlds too often, or for too long, and began to be worried about it, you could try to deny yourself entertainment by telling the system, *"No matter what password I give, don't let me play any more than half an hour of games a day."* This would be a little speed bump [on the infobahn], a warning to slow your involvement with something you found too appealing.' But Mr Gates confesses that, 'Frankly, I'm not too concerned about the world whiling away its hours on the information highway. At worse, I expect, it will be like playing video games or gambling.'[24]

One can imagine a man like Václav Havel breaking into a cracked smile. Havel knows all about both theatrical and real-life gambles. A talented playwright and the founding spokesman of Charter 77, a banned human-rights movement, he actively resisted the oppressions of Communist power in the former Czechoslovakia. His commitment cost him almost a decade of imprisonment and hard labor back in the 1970s and 1980s. These experiences left him with the conviction that there is a universal tension between what he calls the order of Being and that of extinction.[25] Our obsessive belief in technology belongs to the latter, in his view. He perceives a tragic pathology in the way 'We are looking for new scientific recipes, new ideologies, new control systems, new institutions and instruments. We treat the fatal consequences of technology as though they were a technical defect that could be remedied by technology alone. *We are looking for an objective way out of a crisis of objectivism.* Everything would seem to suggest that this is not the way to go.'[26] For him, the answers lie in changed attitudes rather than new techniques: a sense of realism about what our tools can actually achieve.

The 'objectivity' that is supposedly the glory of binary machines is nothing but a philosophical sham. The enterprises of wired society are anything but neutral. They project a distinct and identifiable value system that combines ethical relativism with a clear partisan choice in favor of a raw, expansionist mode of life. It is a powerful combination, and one that throws up unsustainable extremes that the world should resist. The real challenge is to construct an alternative vision of 'progress' that is based on more socially and ecologically acceptable forms of growth. We need to evolve, as Havel puts it, a 'postmodern face.'[27]

How and where to begin?

For one thing, the politician might strive to become 'someone who trusts not only a scientific representation and analysis of the world, but also the world itself. He must believe not only in sociological statistics, but also in real people. He must trust not only an objective interpretation of reality, but also his own soul; not only an adopted ideology, but also his own thoughts; not only the summary reports he receives each morning, but also his own feeling.'[28] 'Neutrality' must give way to choice and *judgment*.

Over the last four hundred years, the continual outward expansion of Western culture has been motivated by a wish to escape the restraints imposed by natural circumstance. The Europe of the fifteenth century that produced Christopher Columbus was also a Europe of plague, violence, and famine: one whose known mineral resources had been depleted and whose verdant forests were largely felled. Kirkpatrick Sale writes of how Columbus's men 'knew, each of them, that for decades, perhaps centuries, mariners and geographers and travelers had told of places of fantastic wealth out there on the further edge of the ocean, and there was every reason to suppose that those stories of golden cities and magical fountains and fist-sized jewels might well come true for them, and to hope that they might return [home] with riches beyond a grandee's dreams.'[29] Centuries later, similar motivations drove America's frontiersmen toward the West. When the country subsequently bumped up against limits to *resource*-based economic growth, it built a new industrial economy based on Taylorism: the standardization, centralization, and rationalization of every production mode.*

Since then, communications technologies have progressively widened our freedom of physical and intellectual mobility, increased the potential

*Named after Frederick W. Taylor (1856–1915), the US engineer who is considered the father of scientific management theory and whose ideas powerfully shaped the evolution of mass production.

for rationalization, and extended the potential scope of our reach, both as individuals and as organizations. A sufficiently independent 'knowledge worker' can escape with his or her laptop to the beach (although more often teleworking means toil from home to save one's employer the cost of providing desk space). In principle, you can spend whole months in the country, staying in touch with your colleagues via modem and PC. Yet this cosmopolitan 'freedom' that we enjoy, which is now being so radically extended, also carries a high price.

While writing this book, I rented an old farmhouse set high on a hill in the Ardèche, an undeveloped region in the south of France. My aim was to get a feel for what a fully Net-based professional life might be like. The setting was magnificent, with vistas of rolling hills and gorgeous skies as far as you could see. The Ardèche remains one of western Europe's last stretches of untamed country, a region where the wild boar scratch up against your front door in the dead of night and where eagles soar and deer silently glide by during the day. But, for all the beauty of the place and its people, I found it an odd and unsettling experience to simultaneously live among them and also on the Net. I remember the afternoon when a neighbor passed by with his pipe and his old dog, gazed suspiciously at my laptop and all the wires that trailed like an umbilical cord from the terrace back into my house, and wordlessly shook his head and passed on. I remember how my looking from the screen to the sky produced a strange shift in perspective, so stark was the contrast between these vast physical spaces, on the one hand, and the labyrinthine electronic regions where I scoured libraries and databases, exchanged flurries of e-mail, and floated draft theories with distant mind-travelers.

What emerged was an unexpected sense of personal ambivalence about the Net. I felt that my cultural distance from all of these farmers and shepherds had somehow been heightened and exaggerated, even though I lived in their midst. Their lives were made of seasons, of animal-shearing, of shared pastures, of milking and the slaughter. My livelihood consisted of arranging and rearranging insubstantial bits through intangible space. When, at the end of the day, I'd switch off my laptop and set forth into the immensity of this place, sometimes following the high ridge paths, sometimes stopping by the café for a glass of wine, I was disturbed by the detachment I felt – a detachment I hadn't noticed on my last homage to rural France. Then I had experienced the village as a place of social continuities and organic shapes of weathered stone and wood; now I was drawn back to my on-line world of rapid shifts, of multiple windows onto abstract space, of angular lines taking shape and vanishing from the screen. The underlying rhythms of my life and those of this place seemed irremediably and uncomfortably out of sync.

Meanwhile, in an ever more networked economy, the terms and conditions of the local herding tradition have been overturned. In this modest town in the Ardèche, the lifestyle and demographics have radically changed. Two decades ago the schoolhouse on the nearby pass drew over forty local kids every day. Now there are virtually no school-age children in sight and the old schoolhouse stands unpainted and forlorn. An entire generation has cleared out. They see no future in this rough style of life; no gain in the production of crops or the herding of sheep. They have gone to Lyons and cities further afield to hunt for the elusive opportunities and lifestyles absorbed from TV. As the harsh winter sets in, and the visiting writer packs up to return to his urban lair, the remaining townsfolk still gather in the meager warmth that glows from the tailgate of a mobile shop that now arrives to sell them vegetables, meat, and bread. (The small shops in town have long since been forced to shutter their doors.) They haven't the resources to point and click their way to a new world. They have lived and worked here all their lives, through many cycles of snow and high summer heat. They have shared occasional triumphs and endured more adversities than an outsider can hope to fully comprehend. And they have evolved a wisdom of their own sort, drawn from a world in which hardships and suffering, like the terrain itself, can only be negotiated but never be escaped.

Pointing and clicking my way through the world, visiting sites, downloading what I need, passing along to the next frame, am I more in touch with the rhythms and sources of a sustainable life? However abstract and 'virtual' one's life becomes, there is no escape from the physical body and its mortality, from the real presence of others, nor from our ultimate dependence upon a finite earth. Mankind is a long way from offloading its consciousness and Being into the Net – or from nourishing itself with digital bits. Still, this never prevents us from trying to play God. It never stops us from adopting that old Cartesian trick of forgetting the flesh-and-blood people who are being left behind (in what the digerati delicately call 'meatspace') as those aboard the wired ark surf their tsunami wave of glittering digital dreams. This special trick, one which all conquerors adopt, 'has unpleasant consequences for those bodies whose speech is silenced by the act of our forgetting ... [indeed] those upon whose labor the act of forgetting [our bodies] is founded ... '[30]

It is often tempting to sentimentalize the old ways. There was never a 'perfect' primeval condition, and I, for one, have no interest in static cultural states. Also, one grows attached to the things one can do with an Apple Mac. One revels in the wonderful sense of freedom that comes with being able to behave differently in fast-shifting states – in short, to have one's cake and eat it too. Like many who work on the Net, I'm thrilled by aspects of my life on-line; here, as the MIT sociologist Sherry Turkle has observed, 'we are encouraged to think of ourselves as fluid, emergent, de-centralized, multiplicitous, flexible, and ever in *process*.'[31] As virtual communities evolve, they provide experiences that can enrich us in RL (Net-jargon for 'real life'). What's more, for those who often stop work late at night when the city all around has already gone to sleep, it is pleasant to have 'social areas' in which to visit with friends and 'interact.' And what about those intriguing twenty-four-hour masked balls called 'MUDs' – multi-user dungeons – where one can invent an entirely fictitious identity? If you want structured games like chess or Go, there is always someone (or something) out there waiting to play. You can enthuse about some terrific film, consult fellow cyclists on the best routes for a vacation, develop agendas for political change. The Net is exploding into constellations of vibrant electronic subcultures. You can visit popular web sites like the titillating alt.sex.binaries, the occasionally useful gopher://marvel.loc.gov/11/research/loc, and the other-worldly cgvc2.html. Their explicit lack of geographical substance and historical continuity is reflected in their terrifically convoluted names.

These 'meeting-grounds' share characteristics unique to the net-work. It is too early to assess their effects on society. Nor is it realistic to suggest what instincts might be unleashed amidst the explosive diaspora away from a shared perceptual continuum in time and space toward the infinitely fragmented multiplicities of the Web; we can note, however, that this explosion is unprece-dented in the history of human affairs. The Main Streets, the Trafalgar Squares, and the high mountain passes of the physical world, which grew out of tracks and crossroads where people tangibly met, produced unique commitments and codes of their own. Will they fully surrender to Net 'locations' that one 'visits' to satisfy instant predefined objectives and interact in specific ways – places where anything disagreeable can be zapped out of sight? Surely not. Nevertheless, it seems that this wonderful freedom that comes when experience is a *chosen collage*, rather than an imposition of some small-village values or an inescapable physical terrain, is here to stay – at least so long as the Web itself can be maintained.

But is this really – as the 800-page *Internet Complete Reference* guide modestly suggests – 'by far the greatest and most significant achievement in the history of mankind'?[32] A senior research scientist at Bell Labs, Robert Lucky, has famously proposed that 'the more we are interconnected, the better off we are.' (Less famously, he also concedes that 'there is still something to be said for interacting with real people.'[33]) Yet it would seem that the mere installation of physical connections does not in itself ensure that the networks operate in a beneficial way. In fact, as the organic structures of human culture are reconfigured into systemic clusters of consumption, grouped around one corporate 'node' and the next, and as people's purchasing, socialization, and employment habits are increasingly conditioned by electronic credits and other forms of digital inducement, and as the networked world begins radically to undermine the opportunities and lives of those attempting to survive in physical space, the truth may be different indeed.

The great paradox is that these electronic bridges, designed to span geographical and temporal space, can also interpose a terrible buffer between those who wish to communicate. The Internet, in its current embryonic shape, is probably still a force more for good than for ill. But it is only a small part of a wider and far more powerful information space – one whose sprawling topology actually *undermines* that quality of trust that lies at the core of all relationships, families, and communities. This essential glue is produced by cultivating continuous ethical connections, on-line or off. And indeed it strikes me, sitting atop my lordly hill in southern France, that the price I have paid for electronic mobility is in fact a sense of disengagement, of *disconnection*, from the places through which I pass. I grow part of an abstract world in which it is all too easy and painless to act upon glowing presences that I can never feel; one in which my keystrokes can burst the human relativities that prevail in real-world spaces too distant to touch or smell, to join a dispersed network in which the links that tie my acts to actual consequences grow so incomprehensibly dense as to seem almost unreal.

The prominent ecologist Edward Goldsmith has written of how:

> a village or small town must ... be arranged so as to confer on it a feeling of wholeness and oneness. In south-west France, the two neighboring towns of Marmande and Villeneuve-sur-Lo [*sic*] are said to exert very different influences on their inhabitants. The former is

> stretched out along a main road, the latter, an ancient bastille, is built round a central square. Of the two, it is the latter which is known for its spirited community ... The central square is a very important feature, offering a place where the citizens can gather to run their affairs. The Greeks could not conceive a city without its agora. Significantly, in the industrial cities of the West, as economic concerns take over from social ones, it is the shopping precinct with its multi-storied car park that is the focal point.[34]

Maybe one day an audio-visually enhanced Net can be configured to produce such a well-rounded communal space; so far, however, computational imperatives have directly and indirectly fragmented the centers of human cohesion. Instead of townsfolk gathering around pattering fountains in the dappled sunlight of the Haute Provence, consumers visit hypermarkets, which are in turn efficiently plugged into a series of networked nodes. The hypermarket is one slot in a multi-storied urban car park – something engineered to provide nothing more and nothing less than efficient message exchange. As the winners of the networking game deploy their considerable talents and tools to transmute abstract digital bits into literal gold, the town square and the life around it fall through the cracks.

'Connectivity' – the technical aspect of message exchange – can never substitute for *connection* – which is the living and continual dialog shared between people and the landscapes that surround them. If electronic networks are allowed to become the dominant conduits for human interaction, and if the narrow priorities of those who are now seeking to assert control over their architecture are successfully imposed, then the reservoir of inherited continuities, of evolved relativities, and of general civil stability – all fed by those clear wellsprings of shared principle that unite places in all their diversity – may well be drained off in order to satisfy the slash-and-burn expediencies of a distant and unaccountable elite. Already it begins to seem as if the more we 'interact' by digital means, the more we are disconnected from each other, the more we are distanced from the social codes which lend meaning to our lives, the more we are deprived of the richness of direct, face-to-face, exchange with our fellows and friends. The windows to the

soul are slamming shut. Accountability is diminishing with every
added length in the lines of communication. Moorings in principle
are jettisoned as physical distances increase. 'Our communities
grow more fragile, airy, and ephemeral, even as our connections
multiply.'[35] We are cast adrift on electrified tides of increasingly
random and inconsequential noise.

The pace and scale of the transformation which was inaugurated
by the Industrial Revolution, and which is indeed still under way
in many parts of the world today, are being dramatically increased
by electronic means. Binary machines, processing abstract inputs
and outputs, can transform the relativities of human life far faster
than the resulting contradictions can be resolved. The unwired
majority of the world thus has every reason to be suspicious, if
not resentful, of a wired outsider's arrival in their midst. Like the
naturalist in Mario Vargas Llosa's novel, a privileged member of
the wired world might rightly feel he or she has everything in
common with the unwired majority: all of our lives – like the lives
of the Indians before – are being colonized by an amorphous elite
from a distant domain. But, tragically, it is often the divisions and
not the unity that tell. Tribe turns against local tribe in a cycle of
senseless violence – while the well-to-do glide placidly on their
ways. Nevertheless, for now at least, one can still appear in a
remote mountain café and the villagers will neither reject nor
double-click one away: more often, indeed, one will be cautiously
welcomed. Real, physical, presences are hard to avoid: they
provide the incentive for people to make contact – that is, to
converse in the fullness of gesture and countenance and to search
for points of social correspondence. But for how long will this
remain the case? How long can the hope of dialog survive in an
atmosphere of organized violence and sanctioned aggression?
Indeed, has the electronic age even begun its search for a sustaining
ethos – one that will help salve precisely those painful human
fractures that the digital 'revolution' both creates and then purports
to heal by electronic means?

In the 1950s Thomas Kuhn published his now-famous treatise *The
Structure of Scientific Revolutions*. In it he challenged the conventional

idea of 'progress' and suggested that, instead of a cumulative and logical progression toward truth, the scientific establishment essentially adopts a succession of world-views which *seem*, for a convenient interval, to explain reality best. He called these world-views 'paradigms'. He described how they tend to become calcified and rigid with time. Contradictory observations are either excluded or bent into impossible shapes in order to conform. Eventually the pressures build up and undermine the structure's foundations from beneath – or else its old adherents simply die away. Kuhn's analysis has been criticized for being overly simplistic, but it stands as a useful way of interpreting events. Today, for example, we can see how Darwinian theory, which suggests that the earth's species evolved by means of gradual adaptation and the survival of its fittest, is now bumping up against contradictory evidence of an altogether more cataclysmic and accidental dynamic – one involving periodic spurts of creation and mass extinction. Newtonian physics, built around atoms and the movements of energy, has similarly given way to a *cybernetic* proposition that the world can be understood in terms of systems and information exchange.

This cybernetic perspective has spread far beyond the hermetic corridors of science to become the dominant paradigm of our times. It is characterized by a singular obsession with problems of *control*, by a tendency to interpret reality in 'systemic' terms that exclude the spirit of nature and man, by a fragmentary (and fragmenting) zero/one language called binary code, and by the ever-accelerating microchip-driven computer cycles and the larger business systems they fuel. Fundamental attributes of this paradigm have been imprinted, via its technologies, onto the human landscape at large. As Sherry Turkle states, 'we are made up, mind and body, of information.'[36]

The intriguing and ironic fact about this supposedly 'revolutionary' mindscape, however, is that it is being used to project a far more ancient paradigm – one in which aggressive expansionism is equated with *both* survival and material prosperity. Although these two imperatives are now radically at odds, the shape of the future is still being cast from those fiery crucibles where commerce and technology meet, in boardrooms and committee chambers that either have completely lost touch with the guiding energies of a sustainable life or have lost the will and judgment to act sensibly on the information they already possess. Improved tools are still projecting an unimproved and thoroughly unrevolutionary agenda that the cultures and ecology of the world can no longer sustain. Will this paradigm be overthrown in an evolutionary cataclysm – or simply fade away? Will society confront the fundamentally ancient choices that are being posed by its latest spurt of technical invention – and engineer an orderly revolution – or will it fall victim to extremities first?

The burden of human existence is that we are each cut off from the great continuity of mother Nature for a brief trajectory that becomes our individual life. We each move through life, making the choices that define us, trying to fill the gaps as best we may. In all of our individual achievements, however, we remain alone. If we are bound at all, it is only by the languages and values we share. As the universal language and technologies of the information age widen the potential scope of human movement from national to global and to virtual, it is time to question whether they will automatically give rise to a binding language and set of values as well. Is it possible that, instead, they are imposing a binary babble of mutually incomprehensible tongues? When great existential and perceptual spaces divide us – for example, when the cultures of the East meet those of the West; when rich meet poor; when the fragmented communities of the Net meet those built on continuities in nature – the only effective lines of connection are those that contain the truest of codes. Without the old verities of the spirit – 'love and honor and pity and pride and compassion and sacrifice,' as the novelist William Faulkner once wrote – the human narrative is well and truly doomed.[37]

Perhaps communication, in its most transcendent sense, is really about love. It is built on sharing and mutual respect. It requires an accommodation of different states of being and seeing. Yet a cybernetic world of fractal movement – one comprised of sources, transmitters, receivers, destinations, feedback loops, and above all of information and message exchange: a realm bound by a two-digit code – has yet to acknowledge the fundamentally spiritual negotiations that universally lie at the heart of its ways.* The quest for human accountability must be pursued irrespective of whether the landscape has changed. Patterns of

*These elements of communication were first enunciated by Claude Shannon and Warren Weaver in the late 1940s.

irresponsible expansion that thrive on the suppression of local relativities, whether cultural or ecological, must be recognized as unacceptable, outdated, and ruinous. The cult of unmodulated rationality has become a dysfunctional religion. The secular individualism that sprang out of fifteenth-century Europe is now in the throes of death. That is why this is such a dangerous time. The suicidal instincts of endless expansion, which belong to an earlier time when there were fewer of us and the earth seemed without bounds, are now being given a careening last run by technical and deregulatory means.

Those who would own and manage life for their own personal benefit have the force of nature working against them. Life will resist their arrogant claims. The question is how its resistance will take shape, and how creative it will actually prove to be. Humankind has yet to make its next evolutionary leap: to acknowledge that our survival both as a species and as civilized creatures depends on whether we can find the modesty to embrace terrestrial limits, to enter into a more respectful dialog with each other, and to assume a more balanced position on the planet we share. This, indeed, is the central challenge of the decades to come. It is time for the information age to pause amid the hasty onrush of impatient change. It is time to reconsider what progress – and *communication* – really mean.

References

Chapter 1

1 Suárez Miranda, *Viajes de Varones Prudentes*, libro cuarto, cap. XIV (Lérida, 1658), quoted in Jorge Luis Borges and Adolfo Bioy Casares, *Extraordinary Tales*, trans. and ed. Anthony Kerrigan (London: Allison & Busby, 1990), p. 123.

2 Nicholas Negroponte, in the Introduction to his *Being Digital* (New York: Vintage Books, 1995).

3 Author and software specialist Michael Dunkerley in *The Jobless Economy? Computer Technology and the World of Work* (London: Polity Press, 1996). See also Richard Donkins, 'Paradise lost and the Protestant work ethic,' *Financial Times*, 3 May 1996.

4 Bill Gates with Nathan Myhrvold and Peter Rinearson, *The Road Ahead* (London: Viking, 1995).

5 See note 2.

6 See the note on pages 180–81, for further detail on the Internet's structure.

7 Bill Gates interviewed in *De virtuele jungle*, a television documentary aired by the Netherlands' NOS broadcast network on 19 December 1995.

8 Quoted in Steven Jay Gould, *Bully for Brontosaurus: Reflections in Natural History* (New York: W. W. Norton & Co., 1991), p. 327.

Chapter 2

1 Malevich correspondence written in 1920 in the midst of his now-famous Suprematist period, during which he was trying to express an 'entire system of world building.' Quoted by Dmitrii Sarabianov in *Malevich 1878–1935* (Amsterdam: Stedelijk Museum, 1989), p. 70.

2 From an advertisement for the '3DO Experience,' a range of video-game titles produced by Electronic Arts, that appeared in *Wired* (US) 1.06, December 1993.

3 Louis Rosetto, Editor's note: 'Why Wired?', *Wired* (UK) 1.01, April 1995.

4 A RAND publication cited by Adam Curtis in the BBC documentary *To the Brink of Eternity*, aired in 'Pandora's Box,' 31 May 1992.

5 Joseph Kraft, 'RAND: Arsenal for ideas,' *Harper's Magazine*, July 1960.

6 Cited by Adam Curtis in *To the Brink of Eternity*, see note 4.

7 Quoted in 'Special report on planners for the Pentagon,' *Business Week*, 13 July 1963.

8 Sam Cohen recalled his role as one of the Cold Warriors in *To the Brink of Eternity*; see note 4.

9 Václav Havel, 'A call for sacrifice,' *Foreign Affairs*, vol. 73, no. 2, March/April 1994.

10 Václav Havel, 'The end of the modern era,' *New York Times*, 1 March 1992; excerpted from his speech to the World Economic Forum in Davos, Switzerland, on 4 February 1992.

11 See James Gleick, *Chaos: Making a New Science* (New York: Penguin, 1988) and M. Mitchell Waldrop, *Complexity: The Emerging Science at the Edge of Order and Chaos* (New York: Simon & Schuster, 1993).

12 See Sushil Wadhwani and Mustaq Shah, *Emerging Giants, Globalization and Equities*, a portfolio-strategy report published by Goldman Sachs on 19 January 1994. See also *World Investment Report 1993: Transnational Corporations and Integrated International Production* (New York: United Nations Publications, 1993).

13 Louis Rosetto, Editor's note: 'Why Wired?', *Wired* (UK) 1.01, April 1995. (My italics.)

14 Genesis 1:31 – 'And God saw everything he had made and, behold, it was very good.'

15 From an interview in *Icon Earth*, a BBC 'Horizon' documentary produced by David Malone and aired on 20 March 1995. (My italics.)

16 Paul Keegan, 'The digerati!', *New York Times Magazine*, 21 May 1995.

17 From the *Magna Carta*; see the footnote on pages 32–3.

18 A frequent refrain, expressed in this case by a participant in The Annenberg Washington Program's seminar on 'The potential downside of the national information infrastructure,' 4 December 1994. The speaker explicitly likened America's 'superhighways' project to the Cold War mobilization that followed Moscow's launch of the satellite *Sputnik* in the 1950s.

19 US president Bill Clinton, inaugural address, Washington DC, January 1992.
20 *To the Brink of Eternity*; see note 4.
21 Norbert Wiener, *God and Golem, Inc.: A Comment on Certain Points Where Cybernetics Impinges on Religion* (Cambridge, Mass.: The MIT Press, 1964), p. 53.
22 The author Fred Halliday quoted by Edward Mortimer in 'The unshakable grip of war in the South,' *Financial Times*, 5 September 1989.
23 Robert Hughes, *Culture of Complaint: The Fraying of America* (New York: Oxford, 1993), p. 27.
24 Frances Williams, 'Global business a fact of life,' *Financial Times*, 31 August 1994.
25 Andrew Bolger, 'Strength ebbs away from fragmented workforce,' *Financial Times*, 31 January 1995.
26 William Bridges, 'The end of the job,' *Fortune*, 19 September 1994.
27 See Ethan Kapstein, 'Workers and the world economy,' *Foreign Affairs*, vol. 75, no. 3, May/June 1996.
28 From the *Magna Carta*; see the footnote on pages 32–3.
29 Computer Systems Policy Project (CSPP), *Perspectives on the National Information Infrastructure* (Washington DC: CSPP, 1993); see pp. 1–3. The CSPP is an affiliation of chief executive officers of corporations that develop, build, and market information-processing systems and software.
30 *Growth, Competitiveness, Employment: The Challenges and Ways Forward into the 21st Century* (Brussels: The Commission of the European Communities, 1993), pp. 166–7.
31 Stewart Brand, *The Media Lab* (New York: Penguin Books, 1988), p. 211.
32 For an in-depth discussion of the many legal implications of cybernetic technology, see Anne Wells Branscom's book *Who Owns Information?* (New York: Basic Books, 1994).
33 From the *Magna Carta*; see the footnote on pages 32–3. This concept will be explored in depth in later pages.
34 The concept of 'hive mind' is explored at length in Kevin Kelly's book *Out of Control* (Reading, Mass.: Addison-Wesley, 1994).
35 Nicholas Negroponte, *Being Digital* (New York: Vintage Books, 1995), p. 157.
36 C. G. Jung, *Memories, Dreams, Reflections* (New York: Vintage Books, 1989), p. 236.

37 Marshall McLuhan, 'The Playboy Interview,' *Playboy*, March 1969.

38 Edward Mortimer, 'Trapped between despair and reform,' *Financial Times*, 23 March 1995.

39 Ian Buruma, *God's Dust: A Modern Asian Journey* (London: Vintage Books, 1991), p. 94.

Chapter 3

1 Marshall McLuhan, *Understanding Media: The Extensions of Man* (New York: Signet, 2nd edn, 1964), p. 32. (My italics.) (Pasteur discovered that microorganisms cause disease.)

2 *Growth, Competitiveness, Employment: The Challenges and Ways Forward into the 21st Century* (Brussels: The Commission of the European Communities, 1993), p. 121.

3 Louis Rosetto, Editor's note: 'Why Wired?', *Wired* (UK) 1.01, April 1995.

4 See Harold Innis, *Empire and Communications* (Oxford: Clarendon Press, 1950) for insights into the intimate relationship between communications media and the projection of power.

5 Speech by John Sculley, 'Computers, communications and content.' Included in *Ethics, Copyright, and the Bottom Line*, the transcript of the Symposium on Digital Technologies and Professional Photography, 9 August 1991, Eastman Kodak Center for Creative Imaging (Camden, Maine: Eastman Kodak Company, 1992), p. 15.

6 Peter Schwartz interviewed in *De virtuele jungle*, a television documentary aired by the Netherlands' NOS broadcast network on 19 December 1995.

7 G. Dan Hutcheson and Jerry D. Hutcheson, 'Technology and economics in the semiconductor industry,' *Scientific American*, January 1996.

8 T. George Harris, 'The post capitalist executive: An interview with Peter F. Drucker,' *Harvard Business Review*, May–June 1993.

9 Michael Dunkerley, author of *The Jobless Economy? Computer Technology in the World of Work* (London: Polity Press, 1996), interviewed by Richard Donkins in 'Paradise lost and the Protestant work ethic,' *Financial Times*, 3 May 1996. (My italics.)

10 Charles Handy, 'The intellectual organization,' *Financial Times*, 23 December 1993. (My italics.)

11 David Bottoms in 'Cyber-cowboy ... or prophet?', *Industry Week*, vol. 244, no. 22, June 1995.

12 Author interview with Hans Decker, former chairman of Siemens North America and executive in residence at Columbia Business School, University Club, New York City, 2 August 1994.

13 Gary Wolf, The *Wired* interview: 'Steve Jobs: The next insanely great thing,' *Wired* (US) 4.02, February 1996, p. 160.

14 Cyrus Freidheim, in a speech at the Davos World Economic Forum, December 1993.

15 See John Berger, Introduction to *Into their Labours: A Trilogy* (New York: Pantheon Books, 1992), p. xxi.

16 Karen Fossli, 'World seaborne trade at record,' *Financial Times*, 2 February 1995.

17 Andrea Lee, 'Prince of books,' *The New Yorker*, 26 April 1993.

18 See Nicholson Baker, 'Discards', *The New Yorker*, 4 April 1994.

19 Damian Fraser, 'Mexico's open door lets in winds of change,' *Financial Times*, 17 July 1994.

20 Tony Jackson, 'Work harder, or not at all,' *Financial Times*, 14 November 1994. For statistics, see also Andrew Jack, Motoko Rich, and Emiko Terazono, 'The longer daily grind,' *Financial Times*, 18 January 1995.

21 Kevin Kelly, *Out of Control* (Reading, Mass.: Addison-Wesley, 1994).

22 Kevin Kelly interviewed in *De virtuele jungle*, a television documentary aired by the Netherlands' NOS broadcast network on 19 December, 1995. For further information see the chapter entitled 'Hive mind' in Kelly, *Out of Control*.

23 See David Korten, *When Corporations Rule the World* (London: Earthscan, 1995), p. 217 – a book that the *Financial Times* believes, title notwithstanding, 'should rank high on the list of "must-read" books for even the busiest corporate executive.' Korten is a Stanford-trained economist. He has worked as a specialist on several Ford Foundation aid projects, and now runs the People-Centered Development Forum in Washington DC.

Chapter 4

1 Cornelius Tacitus, *The Histories*, Book II, XXXVIII, trans. Clifford Moore (Cambridge, Mass.: Harvard University Press, 1925), p. 223.

2 A report by the Landescentralbank, or regional central bank, of the state of Hesse, as reported by Andrew Fisher in 'Frankfurt urged to catch up with London telecoms,' *Financial Times*, 22 February 1996. The report states that 'banks tend to concentrate business activities

with a high telecommunications requirement in those places best equipped in that respect.'

3 Henning Christophersen, EU economic commissioner, quoted in Hugo Dixon, 'Super-highways sans frontières,' *Financial Times*, 21 February 1994.

4 Author interview with Robert Pepper, director, Office of Plans and Policy, The Federal Communications Commission, Washington DC, 6 January 1993.

5 Martin Dickson, 'AT&T plugs into a new market,' *Financial Times*, 6 November 1992.

6 Author interview with Dr Pieter van Hoogstraten, strategy director, Royal PTT Netherlands, The Hague, 10 August 1994.

7 'Make way for multimedia,' *The Economist*, 16 October 1993.

8 Ibid.

9 Author interview with William Wright, European communications director, EDS, 14 September 1994. Players in the business include IBM's Integrated Systems Solutions, General Motors' EDS subsidiary, and Anderson Consulting. 'Companies like ours become part of the fabric of a company ... privy to their strategy and very much a part of their management team,' Mr Wright says.

10 See Fareed Zakaria, 'Culture is destiny: A conversation with Lee Kuan Yew,' *Foreign Affairs*, vol. 73, no. 2, March/April 1994.

11 Larry Donovan, 'Wired for trade in Singapore,' *Financial Times*, 20 July 1993.

12 Quoted by Louise Kehoe in a combined video/print executive briefing project entitled *Beyond 2000 Business Intelligence Report* (London: Financial Times Television/Bain & Co./Business Week, 1993).

13 Interview with John Landry in his capacity as CEO, Lotus Corporation, in *Beyond 2000* (see note 12), one of a series in which leading industry consultants and chief executives discussed the trendlines shaping the global information landscape.

14 Hugh Aldersey-Williams, 'A better use for design,' *Financial Times*, 1 February 1994, and author conversation with Dr Kerckhove, UNESCO Forum, Cologne, Germany, 6 October 1992. (My italics.)

15 Data extracted from the FT 500, a list of the world's leading corporations, published in *Financial Times*, 20 January 1995.

16 See Louise Kehoe, 'Apple launches assault on Intel,' *Financial Times*, 14 March 1994, and Jim Carlton, 'Microsoft software for new Apple line to be widely available by October,' *Wall Street Journal*, 2 August 1994.

17 See Denise Caruso, 'Microsoft morphs into a media company,' *Wired* (US) 4.06, June 1996.

18 See Brent Schlender, 'What Bill Gates really wants,' *Fortune*, January 1995. Also see Charles Morris and Charles Ferguson's fascinating 'How architecture wins technology wars,' *Harvard Business Review*, March–April 1993: 'Competitive success flows to the company that manages to establish proprietary architectural control over a broad, fast-moving competitive space.'

19 From the editorial, 'Rupert at 65,' *Financial Times*, 12 March 1995.

20 Tim Jackson, 'Brokers tremble as E*Trade takes off,' *Financial Times*, 11 March 1996.

21 Author interview with Guido Rey, Rome, 16 April 1987; see David Brown, 'Economist skeptical of Italy's "economic miracle,"' *International Herald Tribune*, 20 April 1987.

22 See the chapter entitled 'The orders of simulacra' in Jean Baudrillard's *Simulations*, trans. Paul Foss, Paul Patton, and Philip Beitchman (New York: Semiotext(e) Foreign Agents Series, 1983).

23 Author interview with senior Clinton administration official, Washington DC, 21 July 1994.

24 See note 7.

Chapter 5

1 Washington Irving, *The Alhambra* (1832): The Complete Works of Washington Irving, vol. 14, ed. William T. Lenehan and Andrew B. Myers (Boston, Mass.: Twayne, 1983), p. 4.

2 Former German chancellor Helmut Schmidt as reported by Joel Kurtzman in 'Welcome, Mr Matsushita, to the New World,' a speech delivered at the First Electronic Money Colloquium, Columbia University Business School, CITI Institute, 21 April 1995.

3 Robert Reich, 'A hand across the great divide,' *Financial Times*, 6 March 1996.

4 George Graham, 'Clearing bank plan to protect forex deals,' *Financial Times*, 11 March 1996; Philip Gawith, 'Bankers clash over global clearing plan,' *Financial Times*, 27 March 1996; George Graham, 'BIS outlines forex settlement risk strategy,' *Financial Times*, 28 March 1996.

5 Joel Kurtzman, *The Death of Money* (New York: Simon & Schuster, 1993), p. 161.

6 Leslie Crawford, 'High on the danger list in Mexico: Ailing companies

queue up for special government help to avoid bankruptcy,' *Financial Times*, 24 January 1996.

7 See Anthony Sampson, *The Money Lenders: the People and Politics of the World Banking Crisis* (New York: Viking/Penguin, 1983), pp. 101, 21.

8 Walter Wriston, *The Twilight of Sovereignty* (New York: Charles Scribner's Sons, 1992), p. 9.

9 Gilmore is a founder of the Electronic Frontier Foundation (EFF). The quote, already part of Net lore, appears in his home page at http://www.cygnus.com/~gnu/.

10 Wriston, *The Twilight of Sovereignty*, pp. 80, 35. (My italics.)

11 Author conversation with Alain Levy-Lang, Global Panel, Maastricht, Netherlands, 18 November 1994.

12 Peter Schwartz, a conversation with Peter Drucker, 'Post-capitalist,' *Wired* (US) 1.03, July/August 1993.

13 Peter Drucker, *The Post-Capitalist Society* (New York: HarperBusiness, 1994); cited by Schwartz in 'Post-capitalist' (see note 12). (My italics.)

14 Hannah Arendt, *The Human Condition* (Chicago: The University of Chicago Press, 1958), p. 253. (My italics.)

15 Roberto Calasso, *The Ruin of Kasch* (Manchester: Carcanet Press, 1994), p. 53.

16 Manuela Saragosa, 'Technologist v technocrat in Indonesia,' *Financial Times*, 24 March 1995.

17 Laurie Morse, 'Traders condemn plan to tax futures,' *Financial Times*, 2 February 1995.

18 Ibid.

19 Alain Levy-Lang address, Global Panel, Maastricht, Netherlands, 18 November 1994.

20 Charles Batchelor, 'Tanker rules threaten oil shipments to US,' *Financial Times*, 1 November 1994.

21 This issue of tanker safety has been painstakingly documented by Charles Batchelor of the *Financial Times*. See 'Britain in brief' column, 'One in five tankers "fails safety code",' 29 April 1993; 'US "blacklist" worries world shipowners,' 27 June 1994; 'Tanker rules threaten oil shipments to US,' 1 November 1994; 'Fears ease on tough new tanker liability rules,' 6 January 1995; 'Shell to move seamen's contracts to Singapore,' 18 January 1996; and 'Insurers protest at standard of ship crews,' 2 July 1996.

22 Kurtzman, *The Death of Money*, p. 40.

23 Joel Kurtzman in 'Welcome, Mr Matsushita, to the New World'; see note 2.

24 Author conversation with Mike Nelson at the First Electronic Money Colloquium, Columbia University Business School, CITI Institute, 21 April 1995.

25 'So much for the cashless society,' *The Economist*, 26 November 1994. (My italics.)

26 Ernie Brickell, speech to the First Electronic Money Colloquium, Columbia University Business School, CITI Institute, 21 April 1995.

27 'So much for the cashless society,' *The Economist*, 26 November 1994.

28 Kurtzman, *The Death of Money*, and author conversation with Mr Kurtzman, New York, 21 April 1995.

29 Author interview with Wim Duisenberg, Amsterdam, 9 October 1996.

30 Author interview with Dr Denning, Washington DC, 30 June 1995. See also her paper on 'The future of cryptography' in *Internet Security*, May 1995.

31 John Taylor, head of Hewlett-Packard's UK-based European Laboratories, quoted by Alan Cane in 'Wanted: superhighway cops,' *Financial Times*, 12 December 1994.

32 Tim May, 'Crypto anarchy and virtual communities,' *Internet Security*, April 1995.

33 World Bank data and extract from the *World Investment Report 1993: Transnational Corporations and Integrated International Production* (New York: United Nations Publications, 1993) cited in Tony Jackson, 'Multinationals take lead as world economic force,' *Financial Times*, 21 July 1993.

34 See Paul Hawken, *The Ecology of Commerce* (New York: Harper-Collins, 1993).

35 See Francis Fukayama, *Trust: The Social Virtues and the Creation of Prosperity* (London: Hamish Hamilton, 1995).

36 Frederich August von Hayek, *The Fatal Conceit: The Errors of Socialism* (Chicago: The University of Chicago Press, 1988). (My italics.)

37 Comment made in a US Supreme Court decision: *Compañía de Tabacos v. Collector*, 275 US 197, 400 (1904).

38 Author interview with Eli Noam, Columbia University Business School, New York City, 27 June 1996.

Chapter 6

1 William Gibson, *Neuromancer* (New York: Ace Books, 1984), p. 51.

2 Salman Rushdie, *Imaginary Homelands* (London: Granta Books, 1991), p. 280.

3 Samuel P. Huntington, 'The clash of civilizations?', *Foreign Affairs*, vol. 23, no. 3, Summer 1993. He writes, 'A West at the peak of its power confronts non-Wests that increasingly have the desire, the will and the resources to shape the world in non-Western ways.'

4 Kirkpatrick Sale, *The Conquest of Paradise* (London: Hodder & Stoughton, 1991), p. 90.

5 Michael Benedikt, in the Introduction to his *Cyberspace: First Steps* (Cambridge, Mass.: The MIT Press, 1991), p. 3.

6 See Neil Postman, *Amusing Ourselves to Death* (New York: Viking, 1985).

7 See Gregory Staple, *Telegeography and the Explosion of Place* (New York: Columbia Institute for Tele-Information, Working Paper 656, 1993).

8 Javed Jabbar, 'The global city, not the global village,' *InterMedia*, vol. 23, no. 6, November–December 1995, p. 45.

9 Benedikt, *Cyberspace: First Steps*, p. 3.

10 Gibson, *Neuromancer*, p. 51.

11 See Clive Cookson, 'The ultimate privatization,' *Financial Times*, 9 October 1994, and Frances Williams, '"Bio-piracy" under new fire,' *Financial Times*, 30 November 1993.

12 David Spark, 'Genetic patenting opens up a moral minefield,' *Financial Times*, 30 March 1995. See also *Conserving Indigenous Knowledge*, an independent study commissioned by the United Nations Development Programme (New York: UNDP, 1994), which concludes 'the existing intellectual property system ... is biased toward the largest enterprises with the strongest legal departments.'

13 Philip Bereano, 'Patent pending: The race to own DNA,' *Seattle Times*, 20 and 27 August 1995.

14 Ibid.

15 See Philip Bereano's excellent two-part series 'Body and soul: The price of biotech' and 'Patent pending: The race to own DNA,' in the *Seattle Times* of 20 and 27 August 1995. The Salk quote appeared in the second piece. (My italics.)

16 Louise Kehoe and Paul Taylor, 'Battle for the eyeballs,' *Financial Times*, 23–4 November 1996. (My italics.)

17 Bill Gates interviewed in *De virtuele jungle*, a television documentary aired by the Netherlands' NOS broadcast network on 19 December 1995.

18 See Peter F. Drucker, 'Trade lessons from the world economy,' *Foreign Affairs*, vol. 73, no. 1, January/February 1994.

19 For this all-too-brief discussion of ship cloths, I have drawn from my correspondence with Robyn Maxwell, curator of the Department of Asian Art at the National Gallery of Australia; from his *Cultures at Crossroads* (New York: Asia Society, 1993); and from his *Textiles of Southeast Asia: Tradition, Trade and Transformation* (Melbourne: Oxford University Press and the Australian National Gallery, 1990).

20 A phrase coined by the Irish playwright Brian Friel in *Translations* (London: Faber and Faber, 1981), p. 43.

Chapter 7

1 Ngugi wa Thiong'o, *Decolonizing the Mind: The Politics of Language in African Literature* (London: Heinemann, 1986); reprinted in *The Norton Reader: An Anthology of Expository Prose* (New York: W. W. Norton & Co.), pp. 779–80.

2 Bruce Chatwin, *Songlines* (New York: Viking/Penguin, 1987), pp. 2, 57, 70.

3 Leonardo Sciascia, *The Moro Affair / The Mystery of Majorana*, trans. from the Italian by Sacha Rabinovitch (Manchester: Carcanet Press, 1987), p. 173.

4 Heinz Pagels, *The Dreams of Reason* (New York: Bantam Books, 1989), p. 88.

5 Quoted by Jean Baudrillard in *Simulations*, trans. Paul Foss, Paul Patton, and Philip Beitchman (New York: Semiotext(e) Foreign Agents Series, 1983), p. 103. (My italics.)

6 George Gordon, 'Communication' entry, *Encyclopædia Britannica*, 15th edn, p. 1006. He adds, 'Premises and conclusions drawn from the syllogisms according to logical rules may be easily tested in a consistent, scientific manner, as long as all parties communicating share the rational premises employed by the particular system.'

7 See 'Agents of alienation' by Jaron Lanier, which appears in the *Voyager* e-zine on the Net at http://www.voyagerco.com/consider/agents/jaron.html.

8 Baudrillard, *Simulations*, p. 152.

9 Luther Standing Bear, Oglala Sioux chief (died 1939), quoted in *Native American Wisdom*, ed. David Borenicht (Philadelphia: Running Press, 1993), p. 88. See also Luther Standing Bear, *Land of the Spotted Eagle* (*c.* 1933) (Lincoln, Nebr.: University of Nebraska Press, 1978) and *Stories of the Sioux* (*c.* 1934) (Lincoln, Nebr.: University of Nebraska Press, 1988).

10 Václav Havel, *Letters to Olga* (London: Faber and Faber, 1990), p. 273.

11 Edward Said, *Culture and Imperialism* (London: Vintage Books, 1994), p. 408.

12 Hannah Arendt, *The Human Condition* (Chicago: The University of Chicago Press, 1958), pp. 284, 286.

13 Michael Prowse, 'Endangered species,' *Financial Times*, 20 November 1995. (My italics.)

14 Eli Noam, 'Electronics and the dim future of the university,' *Science*, 13 October 1995; excerpts reported in 'Scholar sees traditional university disappearing with digital speed,' *Columbia University Record*, vol. 21, no. 7, 20 October 1995, and on the Internet. He adds, 'to safeguard the credibility of this function requires universities to be vigilant against creeping self-commercialization and self-censorship.' (My italics.)

15 George Orwell, 'Politics and the English language' (1946), reprinted in *The Norton Anthology of English Literature*, 5th edn, vol. 2, ed. M. H. Abrams (New York: W. W. Norton & Co., 1986).

16 See Fred Cate, 'The First Amendment and the international "free flow" of information,' *Virginia Journal of Constitutional Law*, vol. 30, no. 2, Winter 1990.

17 Stewart Brand, *The Media Lab* (New York: Penguin Books, 1988), p. 242.

18 David Gardner, '"Apocalypse soon" warning,' *Financial Times*, 14 October 1993.

19 David Buchan, 'Lights, camera, reaction,' *Financial Times*, 18 September 1993.

20 James Pressley, 'EU stirs up controversy with paper on film industry,' *Wall Street Journal*, 8 April 1994.

21 Edward Mortimer and David Buchan, The FT interview: 'Jacques Toubon / Mind your language,' *Financial Times*, 29 November 1994.

22 David Buchan, 'France steps up battle with the English language,' *Financial Times*, 24 February 1994.

23 John Andrews, 'Culture wars,' *Wired* (UK) 1.01, April 1995.

24 Author interview with Richard Self, a senior trade negotiator at the White House's Office of the Special Trade Representative, Washington DC, 22 July 1994.

25 'US audio-visual industry seeks to woo Europe,' *World Trade News Digest*, 29 September 1994.

26 David Dodwell, 'US opts to bide its time on audio-visual battle,' *Financial Times*, 15 December 1993.

27 This quip was made by Philippe Bouvard, in his column in *France-Soir* on 10 December 1993.

28 Christopher Dunkley, 'Year of the Brass Rat,' *Financial Times*, 27 December 1995.

29 Raymond Snoddy, 'Murdoch cut BBC to please China,' *Financial Times*, 14 June 1994; two years later, when the BBC returned to the Asia-Pacific after securing a place on the Panamsat 2 satellite, Chinese viewers were still unable to receive the signals, ironically enough because the BBC had leased digital capacity for which viewing equipment is unavailable in China.

30 Reported by Emma Tucker: 'Brussels TV proposals upset advertisers,' *Financial Times*, 10 January 1996.

31 Frank Rich, 'The capital gang,' *New York Times*, 13 January 1996.

32 See Jeffrey Abramson, F. Christopher Arterton, and Gary Orren, *The Electronic Commonwealth* (New York: Basic Books, 1988).

33 Frank Rich, 'The capital gang,' *New York Times*, 13 January 1996. In his article, Rich reviews a controversial book that advocates a stronger commitment to public broadcasting, written by James Fallows and entitled *Breaking the News: How the Media Undermine American Democracy* (New York: Pantheon Books, 1996).

34 Editorial: 'No revolution for software,' *Financial Times*, 29 November 1995.

35 George F. Kennan, *Around the Cragged Hill: A Personal and Political Philosophy* (New York: W. W. Norton & Co., 1993), p. 159.

36 Paul Kerr interviewed Krzysztof Kieslowski in 'A revolution that's turned full circle,' *Observer*, 15 May 1994. When Kieslowski died, in March 1996, the event went virtually unnoticed on the Web.

37 Barry Lopez, 'The passing wisdom of the birds,' from his collection *Crossing Open Ground* (New York: Charles Scribner's Sons, 1988), p. 197.

Chapter 8

1 Václav Havel, *Letters to Olga* (London: Faber and Faber, 1990), p. 167.

2 David Kline, 'Infobahn warrior,' *Wired* (US) 2.07, July 1994.

3 Gore Vidal, 'Paul Bowles's stories,' in *United States: Essays 1952–1992* (New York: Random House, 1993), p. 434.

4 See Jean-Marie Guéhenno, *The End of the Nation-State*, trans. from the French original, entitled *La Fin de la démocratie*, by Victoria Elliot (Minneapolis: University of Minnesota Press, 1995), p. 139.

5 Kevin Kelly, *Out of Control* (Reading, Mass.: Addison-Wesley, 1994), p. 469.

6 Howard Rheingold, *Virtual Communities* (Reading, Mass.: Addison-Wesley, 1993), p. 118.

7 See Daniel Burstein and David Kline, 'Is government obsolete?', *Wired* (US) 4.01, January 1996; this useful article makes a strong *economic* case for regulatory stabilization policies, although it steers away from more 'woolly' social and democratic arguments on the understandable but (I believe) narrow assumption that 'commerce . . . is the engine that drives society.'

8 David Marquand, 'The great reckoning,' *Prospect*, July 1996.

9 Author interview with Edzard Reuter aboard a Daimler-Benz executive aircraft *en route* from Stuttgart to Paris, 9 October 1990.

10 Akio Morita, 'Toward a new world economic order,' *Atlantic Monthly*, June 1993, p. 88.

11 Background conversation with senior EU communications policy-maker, Brussels, August 1994.

12 Author conversations with John Garrett at CNRI, Reston, Virginia, 21 July 1994 and 5 July 1995.

13 Author interview with Cyrus Freidheim, vice-chairman Booz-Allen & Hamilton, 27 July 1994, and remarks at the Davos World Economic Forum, December 1993. (My italics.)

14 'Empires of the 21st century?', *Business Week*, 21 February 1994.

15 Editorial: 'The future of democracy,' *The Economist*, 17–23 June 1995.

16 Author conversation with John Perry Barlow, Washington DC, 30 June 1995.

17 A. Mowshowitz, 'Virtual feudalism: A vision of political organization in the information age,' from *Orwell of Athene*, ed. P. H. A. Frissen, A. W. Koers, and I. Th. M. Snellen (Amsterdam: Sdu Juridische & Fiscale Uitgeverij, 1992), p. 294.

18 Neal Stephenson, *The Diamond Age* (London: Viking, 1995).

19 Esther Dyson, 'Intellectual value,' *Wired* (US) 3.07, July 1995.

20 Marshall McLuhan, *Understanding Media: The Extensions of Man* (New York: Signet, 2nd edn, 1964), p. 73.

21 Ngugi wa Thiong'o, *Decolonizing the Mind: The Politics of Language in African Literature* (London: Heinemann, 1986); reprinted in *The Norton Reader: An Anthology of Expository Prose* (New York: W. W. Norton & Co.), p. 781.

22 See the footnote on pages 32–3.

23 See Scott McCartney, 'Free airline "miles" become a potent tool for selling everything,' *Wall Street Journal Europe*, 17 April 1996.

24 Bill Gates with Nathan Myhrvold and Peter Rinearson, *The Road Ahead* (London: Viking, 1995), pp. 171, 174.

25 Ibid., p. 183.

26 See the chapter entitled 'The post information age' in Nicholas Negroponte, *Being Digital* (New York: Vintage Books, 1995).

27 Jean Baudrillard, *The Ecstasy of Communication*, trans. Bernard and Corline Schutze, ed. Sylvère Lotringer (New York: Semiotext(e) Foreign Agents Series, 1988), p. 40.

28 John Mayo at a colloquium held by the Institute of International Education, 75th Anniversary Forum, New York Public Library, New York City, 27 October 1994.

29 Luigi Pirandello, autobiographical sketch in *Le Lettere*, trans. William Murray, as quoted in *Bartlett's Familiar Quotations* (Boston: Little, Brown & Co., 15th edn, 1980), p. 723.

30 Karl Polanyi, *The Great Transformation* (Boston: Beacon Press, 1971), pp. 140, 42.

31 Ibid.

32 Norbert Wiener, *God and Golem, Inc.: A Comment on Certain Points Where Cybernetics Impinges on Religion* (Cambridge, Mass.: The MIT Press, 1964), p. 69.

33 See Daniel Pearl, 'Future schlock,' *Wall Street Journal*, 7 August 1995.

34 See Paul Hirst and Grahame Thompson, *Globalization in Question* (London, Blackwell, 1996).

35 See Winfried Ruigrok and Rob van Tolder, *The Logic of International Restructuring* (London: Routledge, 1996), chapter 7 ('The myth of the "global" corporation') and p. 179.

36 Ibid., p. 217.

Chapter 9

1 Norbert Wiener, *God and Golem, Inc.: A Comment on Certain Points Where Cybernetics Impinges on Religion* (Cambridge, Mass.: The MIT Press, 1964), pp. 59–60.

2 See *Business Continuity v. Disaster Recovery in the Changing World of Information Technology*, a management white paper published by the Yankee Group, Boston, Mass., in its Yankee Watch series, vol. 4, no. 7, June 1994.

3 Peter G. Neumann, *Computer-Related Risks* (Reading, Mass.: Addison-Wesley/ACM Press, 1995), pp. 295, 297–8.

4 From Robert Oppenheimer's speech to the Association of Los Alamos Scientists, 2 November 1945, reproduced in *Robert Oppenheimer: Letters & Reflections*, ed. Alice Kimball Smith and Charles Weiner (Cambridge, Mass.: Harvard University Press, 1980), pp. 315–24 (p. 324).

5 Norbert Wiener, *Cybernetics* (New York: Wiley, 2nd edn, 1961), p. 177.

6 Quoted in Stanislaw Ulam, *Adventures of a Mathematician* (New York: Scribners, 1976), p. 217, and Steve J. Heims, *John von Neumann and Norbert Wiener: From Mathematics to the Technologies of Life and Death* (Cambridge, Mass.: The MIT Press, 1980), p. 247.

7 John von Neumann, 'Can we survive technology?', *Fortune*, June 1955.

8 James Gleick, *Genius* (New York: Pantheon Books, 1992), p. 182.

9 Louise Kehoe, 'Driving down a "superhighway",' *Financial Times*, 19 November 1992.

10 Norbert Wiener, *The Human Use of Human Beings* (Garden City, NY: Doubleday, 2nd edn, 1954); cited in *Bartlett's Familiar Quotations* (Boston: Little, Brown, 15th edn, 1980), pp. 831–2.

11 Christian Tyler interview with Joseph Rotblat, 'A life spent worrying over the world's problems,' *Financial Times*, 12 January 1996.

12 See Robert Dorsett, 'Risks in aviation,' *Communications of the ACM*, vol. 37, no. 1, January 1994. Dorsett has written the code for a Boeing 727 systems simulator.

13 Ibid.

14 Wiener, *Cybernetics*, p. 176.

15 Author conversation with Karlheinz Stockhausen, UNESCO Forum, Cologne, Germany, 6 October 1992.

16 Quoted in Heims, *John von Neumann and Norbert Wiener*, p. 341.

17 Author conversation with Charles Dollar, National Archives, Washington DC, 31 December 1992.

18 Speech by John Sculley. Included in *Ethics, Copyright, and the Bottom Line*, the transcript of the Symposium on Digital Technologies and Professional Photography, 9 August 1991, Eastman Kodak Center for Creative Imaging (Camden, Maine: Eastman Kodak Company, 1992), p. 20.

19 Alan C. Kay, 'Computers, networks and education,' *Scientific American*, September 1991.

20 Author interview with Jacqueline Hess, Washington DC, 31 December 1992.

21 Kevin Kelly, *Out of Control* (Reading, Mass.: Addison-Wesley, 1994), pp. 125–7. (My italics.)

22 Bill Gates with Nathan Myhrvold and Peter Rinearson, *The Road Ahead* (London: Viking/Penguin, 1995), p. 267.

23 Ibid., p. 269.

24 See 'http://www.netscape.com/newsref/std/cookie_spec.html' on the Web. (My italics.)

25 'In short' column, *Information Week*, 22 May 1995. See also RISKS 17.13 and 17.14 (May 1995), 17.16 (June 1995), 17.21 (July 1995), 17.27 (August 1995), and 17.60 (January 1996) for related postings.

26 From a thread headed 'Warning on using Win95' in RISKS 17.21, July 1995.

27 One contributor to the Edupage (31 December 1995) reports that a syndicated columnist named Gina Smith, writing in the December 1995 issue of *Popular Science*, has predicted a proliferation of computer 'spy' viruses similar to Microsoft Windows 95's registration wizard. Their function is to explore your CPU and determine whether you've legally registered all software you have installed. As Ms Smith notes, 'It's already possible to do this sort of scanning without alerting the user, so it doesn't take much of a futurist to imagine the same sort of stealth technology being used on unknowing bulletin board and Internet users. In fact, I think a trend away from juvenile-prank computer viruses to information-seeking "spy" viruses isn't merely likely, it's inevitable.'

28 Gates, *The Road Ahead*, p. 270.

29 Mark Weiser, 'The computer for the 21st century,' *Scientific American*, September 1991.

30 From a thread headed 'Digital unreality' in RISKS 18.21, June 1996.

31 Speech by Fred Ritchin. Included in *Ethics, Copyright, and the Bottom Line*, the transcript of the Symposium on Digital Technologies and Professional Photography, 9 August 1991, Eastman Kodak Center for Creative Imaging (Camden, Maine: Eastman Kodak Company, 1992), p. 31.

32 Ibid.

33 A. N. Whitehead, *Introduction to Mathematics* (New York: Holt, 1911); cited in *Bartlett's Familiar Quotations* (Boston: Little, Brown, 15th edn, 1980), p. 697.

34 John Hockenberry, 'Pentium and our crisis of faith,' *New York Times*, 28 December 1994.

35 John Barlow, 'Is there a cyberspace?', posted on the Internet at http://www.eff.org/pub/Publications/John_Perry_Barlow/HTML/utne_community.html.

Chapter 10

1 Italo Calvino, *Six Memos to the Next Millennium* (Cambridge, Mass.: Harvard University Press, 1988), p. 124.

2 Mario Vargas Llosa, *Death in the Andes* (New York: Farrar, Straus & Giroux, 1996), p. 99.

3 See Martin van Creveld, *The Transformation of War* (New York: Free Press, 1991).

4 Italo Calvino, 'Cybernetics and ghosts,' in *The Literature Machine: Essays* (London: Secker & Warburg, 1987), p. 19.

5 Michael Heim, 'The erotic ontology of cyberspace,' in *Cyberspace: First Steps*, ed. Michael Benedikt (Cambridge, Mass.: The MIT Press, 1991), p. 77.

6 Bill Gates with Nathan Myhrvold and Peter Rinearson, *The Road Ahead* (London: Viking/Penguin, 1995), p. 271.

7 Ibid., pp. 250–1.

8 Ibid., p. 251.

9 Ibid., pp. 251, 258.

10 Ibid., p. 261.

11 Nicholas Negroponte, in the Introduction to his *Being Digital* (New York: Alfred A. Knopf, 1995).

12 Ibid.

13 *Report by the President's Council for Sustainable Development* (Washington: USGPO, March 1996); see especially data provided by the US Department of the Interior, Bureau of Mines, Branch of Minerals.

14 See Richard Leakey and Roger Lewin, *The Sixth Extinction* (London: Weidenfeld & Nicolson, 1996); the authors conclude that by destroying the biological diversity that is crucial to the maintenance of all life forms, mankind may already have precipitated its own extinction.

15 William Rees and Mathis Wackernagel, 'Ecological footprints and appropriated carrying capacity: Measuring the natural capital requirements of the human economy,' in *Investing in Natural Capital: The Ecological Economics Approach to Sustainability*, ed. A.-M. Jannson, M. Hammer, C. Folke, and R. Costanza (Washington DC: Island Press,

1994); cited by David Korten in *When Corporations Rule the World* (London: Earthscan, 1995), p. 28.

16 Lester R. Brown and Hall Kane, *Full House: Reassessing the Earth's Population Carrying Capacity* (Washington DC: Earthscan, 1995), pp. 60–61. In his chapter 'Entering a new era,' p. 30, Brown cites a recent joint study by the US National Academy of Sciences and the Royal Society of London, *Population Growth, Resource Consumption, and a Sustainable World* (London and Washington DC, 1992), which began, 'If current predictions of population growth prove accurate and patterns of human activity on the planet remain unchanged, science and technology may not be able to prevent either irreversible degradation of the environment or continued poverty for much of the world.'

17 Korten, *When Corporations Rule the World*, p. 38.

18 Ibid., p. 11.

19 John Gray, *Beyond the New Right* (London: Routledge, 1993), pp. 127, 146, 144.

20 Quoted by Robert Taylor in 'Bigger role for governments will be in fashion next year,' *Financial Times*, 28 June 1996.

21 Quoted in Kevin Kelly, *Out of Control* (Reading, Mass.: Addison-Wesley, 1994), p. 126.

22 Ibid., p. 28. (My italics.)

23 Dr Marcus B. Viermenhouk, interviewed by David Williams in 'The human micro-organism as fungus,' *Wired* (US) 4.04, April 1996.

24 Gates, *The Road Ahead*, p. 264. (My italics.)

25 Václav Havel, *Letters to Olga* (London: Faber and Faber, 1990), p. 186.

26 Václav Havel, 'The end of the modern era,' *New York Times*, 1 March 1992; excerpted from his speech to the World Economic Forum in Davos, Switzerland, on 4 February 1992. (My emphasis.)

27 Ibid.

28 Ibid.

29 Kirkpatrick Sale, *The Conquest of Paradise* (London: Hodder & Stoughton, 1991), p. 14.

30 Allucquere Rosanne Stone, 'Will the real body please stand up?: Boundary stories about virtual cultures,' in Benedikt, *Cyberspace: First Steps*, p. 111.

31 Sherry Turkle, *Life on the Screen* (New York: Simon & Schuster, 1996), pp. 263–4. (My italics.)

32 Harley Hahn, *Harley Hahn's The Internet Complete Reference* (Berkeley, Cal.: Osborne McGraw-Hill, 1996).

33 Robert W. Lucky, 'In a very short time,' in *Technology 2001: The Future of Computing and Communications* (Cambridge, Mass.: The MIT Press, 1991), pp. 365–6.

34 Edward Goldsmith, *The Way: An Ecological World View* (London: Ritter, 1992), p. 373, cited by John Gray in *Beyond the New Right*, p. 158.

35 Michael Heim, 'The erotic ontology of cyberspace,' in Benedikt, *Cyberspace: First Steps*, pp. 76, 74.

36 Turkle, *Life on the Screen*, p. 265.

37 William Faulkner, Nobel Prize speech, Stockholm, 10 December 1950.

Index

The author's electronic coordinates are:
dbrown@euronet.nl